CONTENTS

March 18, 2004

Appendix 1: Answers to Post-Hearing Questions

Appendix 2: Additional Material for the Record

NASA–DEPARTMENT OF DEFENSE COOPERATION IN SPACE TRANSPORTATION

HEARING

BEFORE THE

SUBCOMMITTEE ON SPACE AND AERONAUTICS

COMMITTEE ON SCIENCE

HOUSE OF REPRESENTATIVES

ONE HUNDRED EIGHTH CONGRESS

SECOND SESSION

MARCH 18, 2004

Serial No. 108–49

Printed for the use of the Committee on Science

Available via the World Wide Web: http://www.house.gov/science

U.S. GOVERNMENT PRINTING OFFICE

92–514PS WASHINGTON : 2004

For sale by the Superintendent of Documents, U.S. Government Printing Office
Internet: bookstore.gpo.gov Phone: toll free (866) 512–1800; DC area (202) 512–1800
Fax: (202) 512–2250 Mail: Stop SSOP, Washington, DC 20402–0001

COMMITTEE ON SCIENCE

HON. SHERWOOD L. BOEHLERT, New York, *Chairman*

RALPH M. HALL, Texas
LAMAR S. SMITH, Texas
CURT WELDON, Pennsylvania
DANA ROHRABACHER, California
KEN CALVERT, California
NICK SMITH, Michigan
ROSCOE G. BARTLETT, Maryland
VERNON J. EHLERS, Michigan
GIL GUTKNECHT, Minnesota
GEORGE R. NETHERCUTT, JR., Washington
FRANK D. LUCAS, Oklahoma
JUDY BIGGERT, Illinois
WAYNE T. GILCHREST, Maryland
W. TODD AKIN, Missouri
TIMOTHY V. JOHNSON, Illinois
MELISSA A. HART, Pennsylvania
J. RANDY FORBES, Virginia
PHIL GINGREY, Georgia
ROB BISHOP, Utah
MICHAEL C. BURGESS, Texas
JO BONNER, Alabama
TOM FEENEY, Florida
RANDY NEUGEBAUER, Texas
VACANCY

BART GORDON, Tennessee
JERRY F. COSTELLO, Illinois
EDDIE BERNICE JOHNSON, Texas
LYNN C. WOOLSEY, California
NICK LAMPSON, Texas
JOHN B. LARSON, Connecticut
MARK UDALL, Colorado
DAVID WU, Oregon
MICHAEL M. HONDA, California
BRAD MILLER, North Carolina
LINCOLN DAVIS, Tennessee
SHEILA JACKSON LEE, Texas
ZOE LOFGREN, California
BRAD SHERMAN, California
BRIAN BAIRD, Washington
DENNIS MOORE, Kansas
ANTHONY D. WEINER, New York
JIM MATHESON, Utah
DENNIS A. CARDOZA, California
VACANCY
VACANCY
VACANCY

SUBCOMMITTEE ON SPACE AND AERONAUTICS

DANA ROHRABACHER, California, *Chairman*

RALPH M. HALL, Texas
LAMAR S. SMITH, Texas
CURT WELDON, Pennsylvania
KEN CALVERT, California
ROSCOE G. BARTLETT, Maryland
GEORGE R. NETHERCUTT, JR., Washington
FRANK D. LUCAS, Oklahoma
J. RANDY FORBES, Virginia
ROB BISHOP, Utah
MICHAEL BURGESS, Texas
JO BONNER, Alabama
TOM FEENEY, Florida
VACANCY
SHERWOOD L. BOEHLERT, New York

NICK LAMPSON, Texas
JOHN B. LARSON, Connecticut
MARK UDALL, Colorado
DAVID WU, Oregon
EDDIE BERNICE JOHNSON, Texas
SHEILA JACKSON LEE, Texas
BRAD SHERMAN, California
DENNIS MOORE, Kansas
ANTHONY D. WEINER, New York
VACANCY
VACANCY
VACANCY
BART GORDON, Tennessee

BILL ADKINS *Subcommittee Staff Director*
ED FEDDEMAN *Professional Staff Member*
RUBEN VAN MITCHELL *Professional Staff Member*
KEN MONROE *Professional Staff Member*
CHRIS SHANK *Professional Staff Member*
RICHARD OBERMANN *Democratic Professional Staff Member*
TOM HAMMOND *Staff Assistant*

NASA–DEPARTMENT OF DEFENSE COOPERATION IN SPACE TRANSPORTATION

THURSDAY, MARCH 18, 2004

HOUSE OF REPRESENTATIVES,
SUBCOMMITTEE ON SPACE AND AERONAUTICS,
COMMITTEE ON SCIENCE,
Washington, DC.

The Subcommittee met, pursuant to call, at 1:05 p.m., in Room 2318 of the Rayburn House Office Building, Hon. Dana Rohrabacher [Chairman of the Subcommittee] presiding.

COMMITTEE ON SCIENCE
U.S. HOUSE OF REPRESENTATIVES
SUBCOMMITTEE ON SPACE AND AERONAUTICS
WASHINGTON, DC 20515

Hearing on

NASA-DOD Cooperation in Space Transportation

Thursday, March 18, 2004
1:00 p.m.
2318 Rayburn House Office Building

WITNESS LIST

Rear Admiral (Ret.) Craig Steidle
Associate Administrator
Office of Space Exploration Systems
National Aeronautic and Space Administration
Accompanied by
Ms. Karen Poniatowski
Assistant Associate Administrator
Launch Services
National Aeronautics and Space Administration

Major General (Ret.) Robert Dickman
Deputy for Military Space
Office of the Under Secretary of the Air Force
Department of Defense

The Honorable Ron Sega
Director
Defense Research and Engineering
Department of Defense

Mr. Elon Musk
Chief Executive Officer
Space Exploration Technologies

Section 210 of the Congressional Accountability Act of 1995 applies the rights and protections covered under the Americans with Disabilities Act of 1990 to the United States Congress. Accordingly, the Committee on Science strives to accommodate/meet the needs of those requiring special assistance. If you need special accommodation, please contact the Committee on Science in advance of the scheduled event (3 days requested) at (202) 225-6371 or FAX (202) 225-0891.
Should you need Committee materials in alternative formats, please contact the Committee as noted above.

HEARING CHARTER

SUBCOMMITTEE ON SPACE AND AERONAUTICS
COMMITTEE ON SCIENCE
U.S. HOUSE OF REPRESENTATIVES

NASA–Department of Defense Cooperation in Space Transportation

THURSDAY, MARCH 18, 2004
1:00 P.M.–3:00 P.M.
2318 RAYBURN HOUSE OFFICE BUILDING

I. Purpose

The House Subcommittee on Space and Aeronautics will hold a hearing entitled *NASA–DOD Cooperation in Space Transportation* on Thursday, March 18, 2004, at 1:00 p.m. in room 2318 of the Rayburn House Office Building.

The Department of Defense (DOD) and the National Aeronautics and Space Administration (NASA) both depend on rockets manufactured by private sector contractors to launch payloads into orbit—payloads such as reconnaissance satellites, weather satellites, or scientific instruments that are necessary for national security or to carry out research in space. This hearing will explore whether better coordination between NASA and DOD in developing and purchasing rockets could increase the reliability and lower the cost of launch vehicles. The hearing will also explore how DOD and NASA could encourage the emergence of new, entrepreneurial companies that can launch payloads into space.

The hearing will explore the following questions:

(1) To what extent can NASA and the DOD benefit from greater cooperation in the development and purchasing of launch vehicles?

(2) What steps is NASA taking to collaborate with the DOD in order to realize those benefits?

(3) What areas of launch vehicle development are exclusively the role and responsibility of one agency or the other?

(4) To what extent can NASA and the DOD encourage the growth of the U.S. domestic launch market, including emerging U.S. launch vehicle providers who provide unique capabilities?

II. Witnesses

Rear Admiral (Ret.) Craig Steidle, NASA Associate Administrator for the Office of Exploration Systems, is responsible for developing NASA's new launch vehicles. Prior to joining NASA, RADM Steidle was Vice Commander of Naval Air Systems and Director of the Joint Strike Fighter Program.

Major General (Ret.) Robert Dickman, Deputy for Military Space in the Office of the Under Secretary of the Air Force, manages the planning, programming, and acquisition of Air Force space systems. Maj. Gen. Dickman previously commanded the launch wing at Patrick Air Force Base, Florida.

The Honorable Ron Sega, Director of Defense Research & Engineering, is the chief technical adviser to the Secretary of Defense for all scientific and technical matters, basic and applied research, and advanced technology development. A veteran of two NASA Space Shuttle missions, Dr. Sega also serves as a major general in the Air Force reserves.

Mr. Elon Musk, Chief Executive Officer of Space Exploration Technologies or SpaceX, is developing a new, privately-financed family of launch vehicles intended to reduce the cost and increase the reliability of access to space. Previously, Mr. Musk co-founded and was the largest shareholder of PayPal, a company that developed an internet electronic payment system that was sold for $1.5 billion in October 2002.

III. Brief Overview

- **The Department of Defense (DOD) and the National Aeronautics and Space Administration (NASA) each contract with industry to build the rockets, or launch vehicles, needed to launch each agency's payloads into orbit.** For example, from the 1950s through the 1990s, the DOD funded the development of the Atlas, Delta and Titan families of rockets to lift payloads such as reconnaissance satellites of varying sizes into orbit. Used once and then discarded, these rockets are known as expendable launch vehicles (ELVs). NASA's Apollo program designed the Saturn rocket, which was also expendable, to carry very heavy payloads to the Moon. In the 1970s, NASA developed the Space Shuttle, the world's first and only reusable launch vehicle. (While the Space Shuttle was originally intended to be wholly reusable, the version ultimately built is only partially reusable, as the large, orange-colored external tank is used only once.) The government developed these launch vehicles through contracts with various aerospace contractors, the largest of which today are the Boeing Company and the Lockheed Martin Corporation.
- **The domestic launch industry has suffered economically from the recent decline in demand for commercial launches, making the costs of these rockets more expensive.** In addition to serving the government's launch needs, aerospace companies also serve the commercial launch market. For example, satellite telecommunication companies purchase launches from commercial launch vehicle providers to carry their communications satellites into orbit. However, while the government's demand for launch vehicles from aerospace companies has remained steady, the private sector's demand has dropped precipitously in recent years (due in large part to the use of fiber optics and cellular technologies). This sharp downturn in the commercial launch vehicle market increases the prices that commercial providers charge NASA and the DOD. For the past decade or so, U.S. aerospace companies have also faced increasing competition from foreign launch companies, particularly Arianespace, which is partially owned by European governments.
- **The President's new space exploration initiative will require NASA to use more expendable launch vehicles after 2010, which may provide new opportunities for greater coordination with DOD.** The vision for NASA that the President announced on January 14th calls for NASA to retire the Shuttle after assembling the International Space Station, now targeted for completion by 2010. After that, NASA must decide whether it will develop a new heavy-lift expendable rocket, convert the Shuttle (which is a heavy-lift vehicle) into a configuration designed to carry only cargo, or use or modify existing expendable launch vehicles, which are not capable of launching the heaviest loads. The vision also calls for NASA to develop a new Crew Exploration Vehicle (CEV) to carry humans back to the Moon as early as 2015. Such a vehicle would most likely be lifted into orbit on an expendable launch vehicle. Any existing rocket probably would have to be modified to be rated as safe for humans.
- **NASA and the DOD have had mixed success when collaborating on launching payloads into orbit and on developing new technologies.** Some NASA and DOD collaborations have produced spectacular successes. For example, in 1947 the Bell X–1 experimental vehicle (flown by Chuck Yeager) was operated by the Air Force and designed by NASA's predecessor agency, the National Advisory Committee for Aeronautics. On the other hand, the Space Shuttle itself is an example of a collaboration that did not work out as originally intended. Meeting both NASA and DOD requirements made it more difficult and more costly to design, build and operate the Shuttle. Moreover, eventually the Shuttle proved to be too risky for DOD to use. In 1986, when the entire Shuttle fleet was grounded for 32 months in the wake of the *Challenger* accident, DOD was unable to launch critical national security satellites. Partly as a result, DOD stopped using the Shuttle to launch its national security payloads and turned solely to expendable rockets.
- **New entrants in the domestic launch industry have the potential to lower costs, and increase reliability.** Some relatively new companies are beginning to produce new launch vehicles for the commercial sector and for government. One such company, SpaceX, has said that its goal is to reduce the cost and increase the reliability of launching payloads into space by a factor of ten. DOD awarded SpaceX a contract to launch a research satellite this

May on its new Falcon I rocket. NASA has been unwilling to consider making an award to SpaceX, saying that NASA will only launch on types of rockets that have already had at least one successful launch. However, NASA has recently announced its intent to award a contract to Kistler Aerospace Corporation to demonstrate the company's reusable launch vehicle that someday could carry cargo to the International Space Station. (The contract is contingent on Kistler emerging from bankruptcy.)

- **The White House is preparing to update the government's space transportation policy, which is expected to specify the roles DOD and NASA should play in developing future space launch systems.** In 1994, the Clinton Administration issued a National Space Transportation Policy to delineate the roles DOD and NASA would each play in developing new space launch vehicles. Under the 1994 policy, NASA was to concentrate on developing and demonstrating reusable vehicle technology, while the DOD would focus exclusively on expendable launch vehicles. In 2002, the Bush Administration directed the National Security Council to review this policy due to NASA's failure to develop and demonstrate reusable vehicle technology and the downturn in the commercial, expendable launch vehicle market that affected the government's costs. The release of the Administration's new space transportation policy has been delayed due to the Space Shuttle *Columbia* accident, but it is expected later this year. The new space transportation policy is expected to reflect the Administration's space exploration policy objectives.

IV. Issues

- **What are the benefits and drawbacks of NASA and DOD cooperating on developing and purchasing launch vehicles?** Cooperation between NASA and DOD can lead either to lower costs—or to a proliferation of requirements and higher costs, depending on the situation. Cooperation can also either be an acknowledgement of areas where the two agencies' needs and missions overlap—or an improper merging of distinct missions. Congress and the agencies need to figure out how to decide when cooperation is optimal and when it might be harmful.
- **How can the government better encourage the sustainable growth of the domestic launch industry?** Greater cooperation between NASA and the DOD in developing and purchasing rockets might also benefit the industry by increasing demand for those rockets used by both agencies. A healthy domestic launch industry is important for both NASA and the DOD. But NASA has not yet decided what kinds of launch vehicles it will need for either crew or cargo after it retires the Shuttle and, as mentioned above, cooperation between the two agencies is not always appropriate.
- **How can the government foster the entry of new, innovative launch companies to meet the government's needs?** Both DOD and NASA could benefit from the entry of new companies into the launch vehicle market, especially since such companies promise lower costs and greater reliability. However, using these companies also presents a greater level of risk to the agencies because the companies' technology is unproven. The agencies need to balance the need to encourage emerging companies against the need to carry out agency missions with limited risk.

V. Background

History of NASA and DOD Space Transportation Development Efforts

The DOD funded the development of the Atlas, Delta, and Titan families of ELVs (called expendable because they can only be used once) based on ballistic missile technology from the 1950s–60s. In the 1960s, NASA developed the small Scout rocket and the heavy-lift Saturn rockets, both of which are no longer produced. Today, the Boeing Company manufactures the Delta family of expendable launch vehicles and is part of the Sea Launch joint venture with the Russian/Ukrainian Zenit rocket. Lockheed Martin manufactures the Atlas, Athena, and Titan launch vehicles, and Orbital Sciences Corporation manufactures the smaller Pegasus and Taurus launch vehicles. Both Boeing and Lockheed Martin build portions of NASA's Space Shuttle, and both companies own equal portions of the United Space Alliance (USA), which manages Shuttle operations and maintenance.

During the 1980s and early 1990s, NASA and DOD worked together on an ultimately unsuccessful effort to develop a new reusable launch vehicle to replace the Shuttle, as well as new expendable launch vehicles. These programs failed because of a combination of technical failures and problems with funding. One unsuccessful effort to create a reusable vehicle was the X–30 or National Aerospace Plane project initiated by President Reagan. The project was doomed by insurmountable technical hurdles with hypersonic technology and was also affected by the end of the Cold War, which made moot some of the impetus for the project. At the same time, NASA and DOD initiated expendable launch vehicle programs. Those programs—Advanced Launch System, National Launch System, and Spacelifter—were not sustained by either the agencies or the Congress for long enough to fully develop any new system.

President Clinton issued a National Space Transportation Policy in 1994 that designated lead responsibility for improving expendable launch vehicles to DOD and lead responsibility for upgrading the Space Shuttle and technology development of new reusable launch vehicles to NASA.

The 1994 policy directed NASA to conduct research designed to demonstrate by the year 2000 a rocket engine that could fly to orbit using only a single stage (rather than the multiple-stage rockets that are used today). In response, NASA began two experimental flight test programs in 1995, the X–33 (with Lockheed Martin) and X–34 (with Orbital Sciences). Neither program was able to successfully demonstrate a vehicle, and NASA terminated both programs in March 2001. NASA had spent approximately $1.2 billion on the X–33 and $205 million on the X–34 by the time the programs were canceled. Lockheed Martin said that it had spent $356 million of its own money on the X–33.

At the same time, the 1994 policy directed the DOD to work with industry to modernize or "evolve" the expendable launch vehicle fleet under the Evolved Expendable Launch Vehicle (EELV) program "to reduce costs while improving reliability, operability, responsiveness, and safety." The policy also directed the U.S. Government to meet its future launch needs by purchasing commercial launch services.

In 1995, DOD began funding the development of the latest generation of Delta and Atlas launch vehicles through the EELV program. Under that program, DOD has awarded contracts to Boeing valued at $1.88 billion ($500 million for development plus $1.38 billion for 19 launches) for the Delta IV, and contracts to Lockheed Martin valued at $1.15 billion ($500 million for development plus $650 million for nine launches) for the Atlas V. EELV contracts were awarded to both companies to ensure that DOD would not be forced to rely on a single supplier. Each company has spent about $1 billion of its own money on EELV development. DOD also has a variety of other programs to develop new launch vehicles and vehicle components.

Some low-level cooperation between NASA and DOD on rocket technologies continued even under the 1994 policy, but cooperation began again in earnest around 2000. In the wake of failures in the X–33 and X–34 programs, NASA proposed the Space Launch Initiative, under which it would cooperate with DOD on both reusable and expendable launch technologies.

Economic Landscape for Domestic Launch Industry and Recent Developments

DOD hoped the EELV would be less expensive to purchase than previous launch vehicles. However, that assumed a thriving commercial launch business that would add to the demand for the new rockets. Instead, the demand for commercial launches has plummeted. In 1999, 76 commercial payloads were launched, producing $2.3 billion in launch revenues, while in 2003 only 18 commercial payloads representing $1.2 billion were launched. Furthermore, competition has become more intense even as the number of launches has declined.

Today, both Boeing and Lockheed Martin are seeking to negotiate higher launch prices with DOD and NASA, and the agencies predict that launch costs could increase by 50 percent. DOD's efforts to keep both companies in the launch business were complicated recently when it penalized Boeing after the company was found to have used proprietary information from Lockheed Martin. The penalties included losing awards for several launches and restrictions on bidding for some future launches.

The President's space exploration initiative announced on January 14th would have a significant impact on the launch industry. While NASA does use expendable launch vehicles for some of its current needs, such as Earth science satellites, NASA uses the Space Shuttle (and Russian Soyuz vehicles) to launch humans into space and uses the Space Shuttle and Russian vehicles for related cargo needs. Under the President's proposal, the Shuttle would be retired around 2010. The proposal does not say what NASA will use to take cargo to and from the International Space Station after that time or what will be used to launch payloads to the Moon or other locations. The President proposed developing a new vehicle, called the Crew Exploration Vehicle (CEV), to launch humans after the Shuttle is retired, but NASA has not yet decided what kind of rocket would lift the CEV.

As part of its FY05 budget, NASA has proposed eliminating the Space Launch Initiative as a discrete program. NASA is in the process of deciding which elements of the Space Launch Initiative to retain (in other programs) as relevant to the President's exploration proposal. For example, NASA has already decided to cancel one joint project on advanced rocket engines and to continue a joint project to demonstrate autonomous satellite rendezvous capability.

In addition, the National Security Council is working on an inter-agency effort, begun in 2002, to develop a new space transportation policy. The policy is expected to be released later this year.

Emerging Commercial Launch Providers

Space Exploration Technologies (commonly referred to as SpaceX) is a privately funded company developing a family of launch vehicles called Falcon rockets. SpaceX has said it intends to reduce launch costs ultimately by a factor of ten. The Falcon I launch vehicle is a small rocket priced at $6 million per launch, a significant price savings compared to other comparably-sized rockets. The first launch of the Falcon I rocket, carrying a DOD research satellite, is scheduled for mid-2004.

In addition to Space X, other emerging launch providers include Kistler Aerospace and Universal Space Lines. NASA recently announced that it intends to pay Kistler Aerospace about $227 million to demonstrate that it can carry cargo to and from the International Space Station. This contract is contingent on Kistler successfully emerging from bankruptcy.

NASA has also requested $10 million for FY05 to buy launch services from emerging companies. However, NASA's current launch policy forbids NASA to contract for launch services unless the type of rocket being used has performed at least one successful flight. The policy was put in place in the mid-1990s after several rockets failed on their maiden flights. Those rockets were made by Orbital and CTA, which is no longer in business. The DOD does not have an analogous policy for its research satellites, which is why it is able to use SpaceX's new Falcon I rocket.

VI. The witnesses were asked to respond to the following questions in their testimony before the Subcommittee:

Rear Admiral (Ret.) Craig Steidle, NASA Associate Administrator for the Office of Exploration Systems, was asked to address:

- Are there any specific lessons learned from past NASA–Department of Defense (DOD) joint ventures in space transportation development and operations that NASA is applying to future programs, such as the Crew Exploration Vehicle, human-rated EELV, and heavy-lift launch vehicle?
- What are the benefits and risks to NASA from increased collaboration with the DOD in launch vehicle development and purchases to support human space flight missions and develop the next generation launch technologies?
- What steps is NASA taking to encourage the growth of the U.S. domestic launch market, including emerging commercial launch service providers to support the Space Station and launch research payloads? What risks, if any, is NASA willing to take by relying on these emerging launch providers?

Major General (Ret.) Bob Dickman, Office of the Air Force Under Secretary, was asked to address:

- What are the benefits and risks to the Department of Defense (DOD) from increased collaboration with NASA in launch vehicle development and purchases to support DOD missions?
- What steps is the DOD taking to ensure that it leverages the potential benefits of NASA's investments to improve the capabilities of U.S. launch vehicles?
- What steps is the DOD taking to encourage the growth of the U.S. domestic launch market, including emerging commercial launch service providers to support DOD missions?

Dr. Ron Sega, Defense Research and Engineering, was asked to address:

- What is the status of the Administration's review of U.S. space transportation policy?
- How do NASA and the Department of Defense (DOD) coordinate their broad research portfolios for space launch vehicles? How might the DOD's launch and propulsion research and development activities contribute technologies to NASA initiatives?
- How is the DOD using emerging commercial launch vehicle providers, like SpaceX? What risks, if any, is the DOD taking by relying on these emerging launch providers?

Mr. Elon Musk, Space Exploration Technologies, was asked to address:

- What are the benefits and risks for the U.S. domestic launch industry, including emerging U.S. launch vehicle providers, if NASA and the Department of Defense (DOD) collaborated more in the development and purchases of launch vehicles?
- What specific recommendations would you make for how NASA and the DOD can encourage the healthy growth of the U.S. domestic launch market, especially for emerging commercial launch providers?
- What unique capabilities do emerging launch vehicle providers, like SpaceX, provide to NASA and the DOD?

Chairman ROHRABACHER. I hereby call this meeting of the Space and Aeronautics Subcommittee to order, and without objection, the Chair will be granted authority to recess this committee at any time. Hearing no objection.

At today's hearing, we will examine how well NASA and the Department of Defense collaborate on the development of launch vehicle technology. Our focus will also include how NASA and the Department of Defense can do a better job in encouraging the emergence of entrepreneurial space launch companies. The President's recent announcement on space exploration begins a new chapter in the American space experience. Improving the Nation's launch capability is a critical step in achieving the President's goal of exploring new worlds.

Early in his tenure, the President took the first step in realizing this goal by revisiting space launch policies of the 1990s. These policies drove a wedge between NASA and the Department of Defense where the Department of Defense was limited to improving expendable launch vehicles in terms of design and development. The reusables were—and that technology was to be the responsibility of NASA. Well, and I might add that this Chairman supported that compromise at the time. It seemed like the right thing to do at the time, but compartmentalizing launch vehicle development, however, may have had unintended consequences, the unintended consequences of preventing improvements to the national launch capability. And I believe the President's renewed commitment for discovery and exploration will encourage a more comprehensive and cooperative spirit between NASA and the Department of Defense. As long as it does not duplicate, as long as we are talking about not duplicating efforts, this new approach should work to our country's benefit.

NASA and the Department of Defense also must establish investment strategies that promote innovative ideas from the private sector. Purchasing launch services demonstrates a desire on the part of government to adopt market-based solutions and use market-based and private sector options. The DOD has a long history in giving a fair shot to emerging launch providers like SpaceX. Unfortunately, NASA does not share this track record. NASA should become, hopefully, zealous in its approach to supporting space entrepreneurs. This shouldn't just be the Department of Defense. NASA should be encouraging entrepreneurs in the private sector. Only then can we expect the type of contributions coming from our private industry that American industry is capable of.

Cooperation between NASA and the Department of Defense on technology development is not new and has not always been easy. Indeed, there are plenty of examples that suggest that such an undertaking is difficult, at best, but there are shining examples of success as well. But let us face it, both agencies' approaches to developing space transportation requires, in terms of their requirements, derive from very different cultures and philosophies. We have both the NASA and the Department of Defense, two very different organizations, and bridging the gap between them is going to take strong leadership and a lot of Congressional involvement.

As Chairman, I will vigorously support the joint NASA and Department of Defense space launch initiatives that promise significant benefits, not only to NASA, not only to the Department of Defense, but significant benefits to our country as a whole.

[The prepared statement of Mr. Rohrabacher follows:]

PREPARED STATEMENT OF CHAIRMAN DANA ROHRABACHER

Today's hearing will examine how well NASA and the Department of Defense collaborate on the development of launch vehicle technology. Our focus will also include how NASA and DOD can do a better job in encouraging the emergence of entrepreneurial space launch companies. The President's recent announcement on space exploration begins a new chapter in the American space experience. Improving the Nation's launch capability is a critical step in achieving the President's goal of exploring new worlds.

Early in his tenure, the President took the first step in realizing this goal by revisiting space launch policies of the 1990s. These policies drove a wedge between NASA and DOD, where DOD was limited to improving expendable launch vehicle design and development of reusable launch vehicle technology was the responsibility of NASA. Compartmentalizing launch vehicle development may well have had the unintended consequence of preventing improvements to the national launch capability. I believe the President's renewed commitment for discovery and exploration will encourage a more cooperative spirit between NASA and DOD. As long as it is not duplicative, this new approach should work to our benefit.

NASA and DOD also must establish investment strategies that promote innovative ideas from the private sector. Purchasing launch services demonstrates a desire on the part of government to adopt market-based solutions. DOD has a long history of giving a fair shot to emerging launch providers like SpaceX. Unfortunately, NASA does not share this track record. NASA should also become zealous in its approach in supporting space entrepreneurs. Only then can we expect real process in supporting industry.

Cooperation between NASA and DOD on technology development is not new and has not always been easy. Indeed, there are plenty of examples that suggest such an undertaking is difficult at best, but there are shining examples of success as well. Let's face it, both agencies' approach to developing space transportation requirements derive from very different cultures and philosophies. Bridging this gap is going to take strong leadership.

As Chairman, I will vigorously support joint NASA and DOD space launch initiatives that promise significant benefits to them, as well as the Nation.

Chairman ROHRABACHER. Mr. Lampson, you may now proceed with your opening statement.

Mr. LAMPSON. Thank you, Mr. Chairman. And it is nice to be able to join you in this particular chair for this committee. It is going to be a pleasure working with you and on many of the initiatives that we will be facing, and I look forward to the cooperation, not only that needs to exist between NASA and DOD, but also that needs to exist through all levels of our House of Representatives and our Federal Government.

And so I, too, want to welcome the witnesses today to this hearing, and I look forward to your testimony.

As I just said, the topic of the NASA–DOD cooperation on space transportation is certainly an important one. I hope that our witnesses will be able to provide some useful insights into when such cooperation makes sense as well as when it is, perhaps, inappropriate. So NASA and DOD have had a long history of cooperation across a range of activities, dating back to the early years of the space age. While there have occasionally been difficulties and tensions over the years, I believe that the Nation has benefited from NASA–DOD cooperation.

Space transportation, however, is an area where the record has been mixed. The National Aerospace Plane and the National Launch System were two major joint NASA–DOD initiatives that ultimately wound up being canceled. More recently, the two organizations jointly funded the X–37 space test vehicle until DOD decided that it wasn't enough of a priority to warrant its continued participation. I hope that our witnesses will help us to understand why these previous collaborations failed and how we can ensure that future cooperative space transportation efforts will fare better.

Finally, I think that we need to examine the potential impact of the President's space exploration initiative on NASA–DOD cooperation in space transportation. While there has been some discussion of the role that DOD's expendable launch vehicle programs might play in the initiative, my attention is focused on a different matter. Specifically, in order to fund the President's space initiative, NASA is canceling the Space Launch Initiative, and in particular, the Next Generation Launch Technology Program.

Advanced engine programs are also being terminated, and the funding intended for hypersonics research and development is being diverted to the Exploration Systems budget. Indeed, some of these efforts were supposed to be NASA's contribution to the joint NASA–DOD National Aerospace Initiative that was announced with much fanfare only a few years ago.

I am very concerned that we are eating our technological "seed corn" to make the exploration initiative's budget math work. That doesn't make sense to me. And that is not my definition of an "affordable" exploration initiative.

So clearly, we have a lot to talk about today, and I look forward to hearing from the witnesses, and I yield back my time, Mr. Chairman.

[The prepared statement of Mr. Lampson follows:]

PREPARED STATEMENT OF REPRESENTATIVE NICK LAMPSON

Good afternoon. I'd like to welcome our witnesses to today's hearing, and I look forward to your testimony.

The topic of NASA–DOD cooperation on space transportation is an important one, and I hope that our witnesses will be able to provide some useful insights into when such cooperation makes sense as well as when it is inappropriate. NASA and DOD have had a long history of cooperation across a range of activities, dating back to the early years of the space age. While there have occasionally been difficulties and tensions over the years, I believe that the Nation has benefited from NASA–DOD cooperation.

Space transportation, however, is an area where the record has been mixed. The National Aerospace Plane and the National Launch System were two major joint NASA–DOD initiatives that ultimately wound up being canceled. More recently, the two organizations jointly funded the X–37 space test vehicle until DOD decided that it was not enough of a priority to warrant its continued participation. I hope that our witnesses will help us understand why these previous collaborations failed, and how we can ensure that future cooperative space transportation efforts will fare better.

Finally, I think we need to examine the potential impact of the President's space exploration initiative on NASA–DOD cooperation in space transportation. While there has been some discussion of the role that DOD's expendable launch vehicle programs might play in the initiative, my attention is focused on a different matter. Specifically, in order to fund the President's space initiative, NASA is canceling the Space Launch Initiative-and in particular the Next Generation Launch Technology program.

Advanced engine programs are being terminated, and the funding intended for hypersonics R&D is being diverted to the Exploration Systems budget. Indeed, some of these efforts were supposed to be NASA's contribution to the joint NASA–DOD National Aerospace Initiative that was announced with much fanfare only a few years ago.

I am very concerned that we are eating our technological "seed corn" to make the exploration initiative's budget math work. That doesn't make sense to me. And that's not my definition of an "affordable" exploration initiative.

Well, we have a lot to talk about today, and I look forward to a productive hearing.

Chairman ROHRABACHER. Thank you very much. And it will be a pleasure working with you, as it was with your predecessor, and appreciate your thoughtful statement and the points that you have made.

I would like to take the Chairman's prerogative at this moment to call the attention of the Committee and those present at today's hearing to a very important event that has just happened. On Monday, NASA announced that an asteroid would pass closer to Earth than any other previous asteroid that it has charted. So what we have got here is an asteroid that we didn't know anything about two weeks ago, but now, as I understand, passing within 26,000 miles, is that it, of the Earth. When we talk about cooperation between NASA and the Department of Defense, if there is any example where this might be of benefit to us, it might be in tracking and providing some sort of protection against this type of threat. I certainly would like to hear everybody's ideas on this. If we didn't know this, apparently, NASA has catalogued 600 asteroids so far, which is about—which are a half a mile in diameter or larger. And by 2008, 90 percent of these large asteroids will be charted. But there are still thousands of other smaller asteroids out there. And I believe that we need to take this very seriously. This is not something that—I mean, these small asteroids could take out a city, and if you take a look up at the Moon—now that is not saying that the one that they saw could take out a city, I am not sure—but the fact that we just found it is reason for concern. And take a look at the Moon. All of those craters up there didn't happen because of some space debris from the world from our space program; it happened because there are asteroids and meteorites and objects out there that could threaten the Earth as well.

So with that said, without objection, the opening statements of other Members will be put in the written record unless someone has something they would like to jump in with right now. If not, hearing no objection, your opening statements will be put in the record. And I also ask unanimous consent to insert at the appropriate place in the record the background memorandum prepared by the majority staff for this hearing. Hearing no objections, so ordered.

And we have a distinguished panel of witnesses today, and we appreciate you being with us. I would ask you to, if you could, summarize your written statements to five minutes. And if you can get to the heart of the matter, we will have a better chance for a dialogue right after we come back from this vote. We will have a—we will be able to hear you. So actually, we—actually, we have six recorded votes, so we'll get to two witnesses prior to taking off. We will—the Chair will—the Chair is glad to hear from two witnesses and then to take off and then to come back immediately after the last vote. And we apologize. This is out of our control, just sort of like that asteroid coming right over there that we didn't know anything about. But we did know these votes were coming. That is where we had a—all right.

So first, our opening witness is Major General—no, it is not. Our first witness is Rear Admiral, sorry, Admiral, I almost made you a Major General. Craig Steidle, who is responsible for developing NASA's launch vehicles and Crew Exploration Vehicle. He is new to NASA. I appreciated our visit when he came to my office, but he comes here, perhaps new to NASA, but with a wealth of experience from the Pentagon. And if there is anyone that would have a perspective on how NASA and the Pentagon can cooperate, it would be you, Admiral. You may proceed.

STATEMENT OF REAR ADMIRAL (RET.) CRAIG E. STEIDLE, ASSOCIATE ADMINISTRATOR, OFFICE OF SPACE EXPLORATION SYSTEMS, NATIONAL AERONAUTIC AND SPACE ADMINISTRATION, ACCOMPANIED BY MS. KAREN PONIATOWSKI, ASSISTANT ASSOCIATE ADMINISTRATOR, LAUNCH SERVICES

Rear Admiral STEIDLE. Thank you, sir.

Mr. Chairman, Members of the Committee, I will make this very brief, Sir, I would like to thank you for this opportunity to appear today to discuss NASA's ongoing cooperative activities with the Department of Defense.

On January 14, the President visited NASA Headquarters and announced the Nation's vision for space exploration. In his address, the President presented a vision that is bold and forward-looking, yet practical and responsible, one that explores answers to longstanding questions of importance to science and society, and will develop revolutionary technologies and capabilities for the future, while maintaining good stewardship of the taxpayers' dollars. Key to the good stewardship of taxpayers' dollars is appropriate partnering between NASA and DOD.

At this time, I would first like to introduce Karen Poniatowski. Karen will be joining me at the table when we start the session for questions. She is the Associate Administrator for Launch Services in the Office of Space Flight and has a lot of knowledge on the areas that you are interested in today as well.

Chairman ROHRABACHER. So you brought your institutional memory with you. All right.

Rear Admiral STEIDLE. Yes, sir.

She is a significant—she has been a significant contributor to cooperative efforts with DOD on the Space Shuttle, Space Station, expendable launch vehicles and such, so that is why I asked her to come with me today.

There has been a rich history of collaboration and cooperation between DOD and NASA, and I fully expect this tradition will continue. I have some examples that I hope to provide you with later on.

The Office of Exploration Systems, which I have been privileged to head, has implemented a strategy of communicating with our partners and our stakeholders, and we have begun this, holding our first series of NASA days and industry days in our auditorium, and some of the staff in here have attended those. This provides an opportunity for my leadership team to directly communicate and hold a dialogue with our partners about our plans for implementing the Nation's vision. We are currently in the requirements development phase, and that is well underway. Once the requirements have been defined, future opportunities for collaboration, in cooperation with DOD, will be better understood and pursued, and I intend to continue this communication strategy throughout, as I did before in the Joint Strike Fighter program, and I found it very productive to do that.

The Partnership Council, as you will probably hear from my colleagues, is a multi-agency forum with a diverse membership of renowned leaders: Mr. O'Keefe, Mr. Teets, Admiral Jim Ellis, General Lord, and Mr. Sega at the end of the table. The Council is a mechanism for agencies to frame issues and discuss their individual space mission needs and identify actions to benefit the entire space community, and I understand that the Exploration Systems flowed from the original discussions in that partnership.

Project Constellation, which you have referenced, will develop a new Crew Exploration Vehicle for future crew transport. This vehicle will be developed in stages with the first flight demonstration test in 2008, the first unmanned flight in 2011, and the first crewed flight in 2014. Project Constellation will be discussed at the Partnership Council this spring to ensure that our space partners and DOD are kept fully informed about its implementation and our way forward.

Areas of mutual concern to both NASA and DOD include launch assets. Discussions are underway with DOD on the topic of assured space access, exploring the possibility of human rating and enhancing performance and reliability of launch systems to support the Nation's vision for space exploration, and defining a science and technology strategy that will advance the Nation's ability to meet future launch needs.

As we implement the recommendations of the CAIB, the *Columbia* Accident Investigation Board, we are working with DOD to refine their support for spacecraft operations and members of various DOD organizations and facilities are being employed to develop and validate return-to-flight implementations and approaches. And NASA is also partnering with DOD to respond to both technical and cultural issues outlined in the CAIB report.

NASA and the Department of Defense have a long history of cooperation on services, and we hope to have an opportunity to discuss that with you today. Although we have different missions, we share many of the same issues. The technical challenges are the same as are some of the requirements, working together within partnerships to maximize these unique resources. Through numerous cooperative efforts, the American people can benefit from our joint endeavors in space and on Earth, and I sincerely appreciate the forum in this particular committee for providing us today the opportunity to discuss that.

Thank you, sir.

[The prepared statement of Rear Admiral Steidle follows:]

PREPARED STATEMENT OF REAR ADMIRAL (RET.) CRAIG E. STEIDLE

Mr. Chairman and Members of the Committee, thank you for this opportunity to appear today to discuss NASA's ongoing cooperative activities with the Department of Defense (DOD). Let me begin with a discussion of a key part of that cooperation.

Partnership Council

NASA's relationship with the DOD has been coordinated primarily through the Partnership Council since Mr. O'Keefe began his tenure as the NASA Administrator. During the last 18 months, NASA's role has proven instrumental in the evolution of the Council. Early on, key changes began to occur that had elevated the level of the Council members and changed completely how meetings were run. We now have on track an effective mechanism for cooperation in a variety of areas.

The Partnership Council is a multi-agency forum with a diverse membership that includes Mr. Sean O'Keefe, NASA Administrator; Mr. Peter Teets, the DOD Executive Agent for Space and Director of the National Reconnaissance Office; Admiral James Ellis, the Commander of U.S. Strategic Command; General Lance Lord, the Commander of Air Force Space Command; and Dr. Ronald Sega, the Director of Defense Research & Engineering.

The Agency leaders established the Partnership Council to provide a forum for senior DOD and civil space leaders to meet face-to-face on a regular basis to discuss cross cutting issues relevant to the national space community. The purpose of the Partnership Council is to facilitate communication between the organizations and to identify areas for collaboration and cooperation.

On January 14th, 2004, the President visited NASA Headquarters and announced the Vision for Space Exploration. In his address, the President presented a vision that is bold and forward-thinking, yet practical and responsible—one that explores answers to longstanding questions of importance to science and society and will develop revolutionary technologies and capabilities for the future, while maintaining good stewardship of taxpayer dollars. Key to good stewardship is appropriate partnering between NASA and DOD. The Partnership Council provides us the necessary forum for the strategic communication necessary to turn the vision into reality.

The Council has proven an invaluable mechanism to enable the Agencies that use space assets to discuss their individual mission needs and capabilities in a forum where issues can be framed and appropriate actions assigned to benefit the entire space community.

NASA–DOD Cooperative Activities in Space Transportation

Space launch systems are inextricably woven into the fabric of America's national security. As a result, the ability of the United States to launch critical space assets when and where they are needed is a national security requirement. Civil missions are also dependent on assured access. Currently, access to the International Space Station is dependent solely upon Russian launch capability until the Space Shuttle returns to flight. Accordingly, the Partnership Council routinely discusses launch topics to ensure that agencies partner appropriately in their approach.

Return to safe flight is a driving priority at NASA. It is imperative that we are able to return the Space Shuttle to flight in a safety-driven, expeditious manner. As we implement the recommendations of the *Columbia* Accident Investigation Board (CAIB), NASA is working with DOD to redefine DOD support for space flight operations. Memoranda of Agreement concerning the CAIB recommendations are being reviewed and rewritten at all levels.

In addition to the CAIB activities, there are currently over 400 active agreements between various NASA and DOD organizations. Agreements cover a range of activities between each of the NASA Centers and varied groups within DOD.

Looking to the future, there are new opportunities for collaboration and support between NASA and the DOD. For example, Project Constellation will develop a new Crew Exploration Vehicle for future crew transport. This vehicle will be developed in stages, with the first flight demonstration test in 2008, the first unmanned flight in 2011, and the first crewed flight in 2014. Project Constellation will be discussed at the Partnership Council this spring to ensure that our space partners in DOD are kept fully abreast of the ongoing implementation of our Vision for Space Exploration.

For cargo transport to the International Space Station after 2010, NASA will rely on existing or new commercial cargo transport systems. NASA does not plan to develop new launch vehicle capabilities except where critical NASA needs—such as heavy lift—are not met by commercial or military systems. Discussions are underway with DOD on the topic of assured access to space, exploring the possibility of human rating and enhancing performance and reliability of launch systems to support the Vision for Space Exploration, and defining a Science and Technology strategy that will advance the Nation's ability to meet its future launch needs.

Space Shuttle and Expendable Launch Vehicles

There is a rich history of cooperation with the DOD on the Space Shuttle and Expendable Launch Vehicles. From NASA's early days, we have depended on the DOD to provide launch range facilities and support for the NASA Space Shuttle and expendable launch activities on a reimbursable basis at both the Eastern and Western ranges. NASA represents one of the largest reimbursable customers of the Air Force on the Eastern Range. NASA, in close cooperation with the DOD and Industry, established the Advanced Spaceport Technology Working Group and Advanced Range Technology Working Group. Through these government and industry working groups we are able to identify advanced technologies to ultimately improve the performance and reduce the cost of range operations for all range users.

NASA has also utilized USAF-unique launch support for missions for which commercial capability was not available, such as the Cassini mission that required the performance of the USAF Titan IV. Refurbished Atlas E and Titan II services were provided on a reimbursable basis to support flight of the NOAA Polar Meteorological satellites and the quick response QuikScat mission launched in June 1999. NASA has also conducted shared missions, most recently the Kodiak Star launch on an Athena launch vehicle from Alaska in September 2001 and the February 2000, STS–99 Shuttle Radar Topography Mission (SRTM), which was a joint effort of NASA and the National Geospatial-Intelligence Agency (NGA-formerly NIMA). The data collected provided precise, uniform, 3-dimensional elevation data for roughly 80 percent of the land mass of the Earth.

The Space Shuttle is the only reusable launch vehicle in the world capable of transporting humans to and from space. This capability has made it a workhorse for the United States space program for more than 20 years. The Shuttle has been used for a variety of purposes, ranging from launching, retrieving, and servicing scientific payloads to conducting experiments on behalf of the other NASA enterprises to transporting elements of the International Space Station (ISS) into orbit.

The DOD has flown 11 dedicated missions on the Space Shuttle. NASA is proud to have provided space access to about 270 secondary DOD payloads, some 260 on the Shuttle as mid-deck or cargo bay payloads, and four to the Russian Space Station MIR. They were also one of the earliest users of the International Space Station (ISS). To date, four DOD payloads have used the ISS as a space based research platform. Areas of emphasis for these (and future) payloads include surveillance and weather, space control and situational awareness, satellite subsystems, assured/responsive access to space, and education.

At present, NASA remains focused on safe return-to-flight of the Space Shuttle and successful assembly of the ISS. The return-to-flight (RTF) effort is being guided by "NASA's Implementation Plan for Space Shuttle Return to Flight and Beyond." This plan addresses NASA's approach for implementing each of the recommendations from the CAIB report.

Members of various DOD organizations and facilities are being employed to develop and validate RTF implementation approaches to fulfill the CAIB recommendations. NASA is also partnering with the DOD to respond to both technical and cultural issues outlined in the CAIB report. Some examples include the use of thermal-vacuum facilities at the Arnold Engineering Development Center and Eglin Air Force Base to test design modifications to the External Tank, collaborating with the Navy's Submarine Nuclear Reactors Program, and Submarine Safety Program to enhance NASA's processes for evaluating issues and concerns.

We are also working with the DOD to respond to the CAIB recommendations as they pertain to expendable launch vehicles (ELVs). NASA, the United States Air Force, and the National Reconnaissance Office recently held the 4th Government/Industry ELV Mission Assurance Forum on March 9–10, 2004. This forum was originally established by these agencies to ensure that the ELV lessons learned from the 1998 Presidential Broad Area Review into Launch Failures are not lost and continues to be one of the many forums established to facilitate communication between the government agencies with regards to space transportation.

The Vision for Space Exploration

Discussions have begun with the DOD in support of the Vision for Space Exploration. At this point, we are early in the process of defining requirements for the vision. The goal of this process is to develop documented requirements that are traceable, verifiable and measurable.

Definition of Level 0 requirements for the Crew Exploration Vehicle (CEV) is the responsibility of the NASA Space Architect. NASA anticipates the final set of Level 0 requirements by the end of this month, pending approval by the Joint Strategic Assessment Committee (JSAC). The JSAC is comprised of the following NASA personnel: the Space Architect (Chair), the Chief Scientist (Deputy Chair), the Enterprise Associate Administrators, Institutions Management and key functional offices.

Definition, documentation and management of Level 1 and 2 requirements will be the responsibility of the Office of Exploration Systems. Requirements will be subject to an open and formal review and approval process to be managed by the Office of Exploration Systems. The Office of Exploration Systems will develop necessary companion products, including a Management Plan. Parts of this process are the studies and systems analysis of potential exploration scenarios to be conducted as a means of bounding the requirements trade space and developing meaningful figures of merit to be used in the design and development of the CEV.

A Requirements team, lead by Office of Exploration Systems is focusing on developing these requirements and scenarios. This activity began in February 2004. A rigorous requirements formulation approach will yield Level 1 requirements in early September 2004. At that time, the requirements will be provided to the JSAC for approval. The JSAC will present the requirement to the Executive Council, which is composed of the following NASA personnel: the Deputy Administrator (Chair), the Associate Deputy Administrator for Institutions, the Associate Deputy Administrator for Technical Programs, the Enterprise Associate Administrators, the Chief Engineer, the Safety and Mission Assurance Associate Administrator, the Chief Financial Officer, and the General Council. The Executive Council will have the ultimate approval authority for CEV requirements. This effort will be followed by a solicitation in Fall 2004 for concept development. Level 2 requirements will be baselined in early 2005.

In order to develop safe, reliable, and cost-effective requirements for space launch vehicles to implement NASA's Vision for Space Exploration, it is essential to learn from past and present programs such as the Space Shuttle, the Space Launch Initiative, the Orbital Space Plane (OSP), and Next Generation Launch Technology. The Requirements team, lead by the Office of Exploration Systems, will focus on developing these requirements and scenarios utilizing these lessons learned. Knowledge from OSP will help define the fundamental requirements necessary for developing missions beyond Earth orbit.

The Office of Exploration Systems will work closely with the DOD throughout the requirements process. Once requirements have been defined, future relationships with DOD will be further developed in terms of identifying new areas for collaboration and cooperation.

Conclusion

NASA's Office of Space Flight and the Department of Defense (DOD) have a long history of cooperation on services that range from staffing our astronaut corps to collaborating on numerous space technology projects. Although we have different missions, we share many of the same issues, technical challenges, and requirements. Working together we have formed partnerships to maximize our unique resources. Through numerous cooperative efforts, the American people have benefited by our joint endeavors in space and on the Earth.

I sincerely appreciate the forum that the Committee has provided today, and I look forward to responding to your questions.

BIOGRAPHY FOR CRAIG E. STEIDLE

Adm. Craig E. Steidle is the Associate Administrator for the Office of Exploration Systems. He is the first to hold this position, since the office was created in January 2004.

The Office of Exploration Systems was established to set priorities and direct the identification, development, and validation of exploration systems and related technologies. Users and technologists will work together to enable a balancing of factors between requirements, program schedules and costs leading to future space exploration systems.

Since retiring from the Navy in March 2000, Adm. Steidle served as an independent aerospace consultant. His last assignment was as Chief Aerospace Engineer and Vice Commander, Naval Air Systems Command, which develops, acquires and supports naval aeronautical systems.

Adm. Steidle entered the Navy after graduating with merit from the United States Naval Academy, Annapolis, Md. He trained as an attack pilot, flew carrier night combat missions in North Vietnam; served as a test pilot and test pilot instructor; and commanded the Navy's A–3 weapon systems program. During the 1980's, he deployed on carriers, frigates, and cruisers in the Western Pacific and Indian Ocean. Additionally, he served as manager of the Navy's Aerospace Engineers and as the Special Assistant for Air Combat to the Assistant Secretary of the Navy.

Adm. Steidle commanded the Navy's F/A–18 Program, naval aviation's largest production, research and development program, as well as the largest foreign military sales program. The Secretary of Defense presented Steidle with the Navy's Outstanding Program Manager Award.

Adm. Steidle served as the Director of the Department of Defense (DOD) Joint Advanced Strike Technology Office and was the Director of the Joint Strike Fighter

Program, DOD's largest program. Under his command, the Joint Strike Fighter Program was awarded the David Packard Excellence in Acquisition Award.

Adm. Steidle earned a Master of Science degree in systems management from the University of Southern California and a Master of Science degree in aerospace engineering from Virginia Polytechnic Institute. He is a member of the Society of Experimental Test Pilots and a Fellow of the Royal Aeronautical Society.

His decorations and honors include the Defense Distinguished Service Medal; Navy Distinguished Service Medal; Legion of Merit; Distinguished Flying Cross; Meritorious Service Medal with gold star; Air Medals with bronze star; Navy Commendation Medals; Republic of Vietnam Gallantry Cross; and Joint Meritorious Unit Award.

Chairman ROHRABACHER. Well, thank you very much.

And now we are going to give the General a chance, General Bob Dickman, who manages the planning and programming and acquisition of the Air Force Space Systems at the Pentagon. And he has a great deal of experience in working with NASA and commanded the Air Force launch wing in Florida. And we welcome you today, and you may proceed.

STATEMENT OF MAJOR GENERAL (RET.) ROBERT S. DICKMAN, DEPUTY FOR MILITARY SPACE, OFFICE OF THE UNDER SECRETARY OF THE AIR FORCE, DEPARTMENT OF DEFENSE

Major General DICKMAN. Mr. Chairman and distinguished Members of the Committee, thank you for the opportunity to appear before you today to discuss DOD's close relationship with NASA, and in particular, our cooperative efforts in space launch.

In particular, I am honored to be here with Admiral Steidle and Dr. Sega, and Mr. Musk, and our partner in launch, Ms. Poniatowski. And Karen, I am sorry, I got that—something went wrong.

Your hearing today underscores the importance to DOD and NASA and the commercial sector of working together on the challenges in developing launch systems. A healthy and vigorous cooperation, which takes into account the lessons learned from past ventures, is central to sustaining America's preeminence in space. Mr. Chairman, I share your view that there were unintended consequences from that policy of the mid-1990s, but I would point out that it also led to the two finest expendable launch vehicles flying in the world today, the Delta IV and the Atlas V.

Let me start off by saying that cooperation between NASA and the DOD is nothing new, as Craig has pointed out. Civil and DOD space organizations have worked together since 1958. We share our commitment to excellence in our space endeavors. Sometimes we forget that the Atlas and Delta launch vehicles that were the centerpiece of our own launch capability in the military over the last 15 years were developed by NASA and transitioned to military vehicles. There is no issue, then, of whether or not we should cooperate on launch, rather the question is when does the collaboration make sense.

NASA–DOD collaboration and cooperation in space and in aeronautics makes the most sense when our missions, our requirements, and our technologies are similar. Both NASA and DOD benefit from polling our skills and our limited resources. For instance, there is a clear advantage to joint DOD and civil efforts in basic science and technology in areas such as propulsion, materials, avionics, and other launch technologies.

However, cooperation is inappropriate when it crosses the line from civil towards military activities. Tradition and policy dictate that civil and military space endeavors should remain, and do remain, separate.

On a different level, true joint programs, the field operational systems that involve funding from multiple agencies and require design tradeoffs to achieve a common configuration, have proven both difficult to bring to completion and inordinately expensive. Given these bounds on when DOD and NASA might work together, there are many fruitful areas of ongoing and future collaboration.

As I mentioned a moment ago, in the realm of basic science and technology, the military and civil space communities share many common interests. Collaborative S&T projects, in addition to the areas that I mentioned earlier, include sensors, electronics, power generation, communications, thermal protection systems, structures, test facilities, microsatellite technology, and as you have noted, detection of near-Earth asteroids. NASA's Scramjet flight, now scheduled for March 27, will be an important milestone in our hypersonics road map. Cooperative development is also underway through the Integrated High Payoff Rocket Propulsion Technology Program, a three-phase, 15-year national program to double propulsion capability.

Another major DOD priority is developing operationally responsive space lift. The program to make responsive spacelift a reality is called FALCON, not to be confused with SpaceX's Falcon Launch Vehicle. Our FALCON program is run jointly by the Air Force and DARPA, and current efforts are focused on concept design. Mr. Musk's Falcon is one of the competing designs.

I bring the Committee's attention to the FALCON program, because NASA has played an important part in those efforts. In particular, the Marshall Space Flight Center is continually consulting with Air Force Space Command as we have gone through our year-long operationally responsive spacelift analysis of alternatives.

Our major launch program also has the potential to further NASA's space flight efforts, and again, we have a good history of cooperation. I am speaking, of course, about the Evolved Expendable Launch Vehicle. NASA will use both the EELV boosters, the Delta IV and the Atlas V, to launch intermediate and heavy payloads. For example, in 2005, they plan to launch the Mars Reconnaissance Orbiter on the Atlas V and GOES–N on the Delta IV. In addition, both the Atlas and Delta boosters have the potential to be rated for human flight and modified to meet very heavy lift needs. Should NASA choose to do so, the DOD will work with NASA to facilitate their efforts.

While EELV is a major arrow in our space launch quiver, the DOD must continue to look to the future. In this process, we pay careful attention to encouraging the growth of a domestic launch market to include emerging commercial space launch providers. A very appropriate example, considering Mr. Musk's presence here, is SpaceX's Falcon launch vehicle. Later this year, this new launch vehicle is scheduled to carry a satellite into space, launching from Vandenberg Air Force Base.

Let me close by saying we appreciate the support the Congress and this committee have given to help guide and field vital national space capabilities. We look forward to working with you as we continue to develop, produce, and operate critical space systems, both civil and military, for this great nation.

Mr. Chairman, this concludes my opening remarks, and I look forward to your questions.

[The prepared statement of Major General Dickman follows:]

PREPARED STATEMENT OF ROBERT S. DICKMAN

Introduction

While missions and requirements may not always be common, there will always be obvious synergies that allow a close relationship between the DOD and NASA. Our shared environment is hazardous, and we will continue to rely on each other's experience and continued technical innovation to succeed.

In recent years, the Air Force (AF) and NASA have supported each other in a wide range of activities. The four major areas of cooperation were centered on interagency coordination, science and technology development, space operations, and human space flight activities.

NASA–DOD Interagency Coordination

While NASA and DOD have different primary missions, there is significant overlap in the science and technology (S&T) challenges both organizations face. DOD and NASA are aware of, and recognize this potential for, dual use and therefore, the importance of cooperation. To facilitate this cooperation, several forums are acting to promote collaborative planning.

Partnership Council—Initially established in February 1997 by Air Force Space Command (AFSPC) and NASA, this forum is at the most senior level of the planning process. The Partnership Council, consisting of the Honorable Peter B. Teets, Under Secretary of the Air Force and Director, National Reconnaissance Office (NRO); Mr. Sean O'Keefe, NASA Administrator; Admiral James Ellis, Commander, USSTRATCOM; Dr. Ron Sega, Director Defense Research & Engineering; and General Lance Lord, Commander, AFSPC; is the primary forum for high-level discussions between the community. It is intended to achieve efficiencies, effectiveness, risk reduction, and better understanding of plans and activities in areas of mutual interest, to include S&T.

Monthly Meeting between Deputy for Military Space, Office of the Under Secretary of the Air Force; Director, National Security Space Integration; Director, National Security Space Architect; and NASA Space Architect—These National Security Space (NSS) principals and the NASA Space Architect meet regularly, and as needed, to improve the intermediate planning process and products, and implement opportunities identified by the Partnership Council.

Space Technology Alliance (STA)—The STA was initiated in 1997 among the AF, NRO, and NASA to foster cooperative efforts and improve communications.

Other Planning Activities—NASA participates with the NSS community in conducting the annual NSS Program Assessment that, among other things, identifies interagency S&T cooperation opportunities. NASA also participates in the annual update of the NSS Plan that provides implementation guidance to the NSS community on desired capabilities, including S&T. In addition, NASA is participating with the DOD and the Intelligence Community in developing the Congressionally-directed DOD Space S&T Strategy, which will be complete in the summer of 2004.

There are a number of good examples of cooperation and mutual support between DOD and NASA over our long history of working together. These include launch and range support, communications, flight experiments, and environmental science. However, there is always room for improvement. We recognize this as being in the best interest of the Nation and have therefore taken steps to strengthen our efforts with the recent initiation of monthly planning meetings and the development of the Space S&T Strategy.

There are, however, some important differences between NASA and the NSS community—one open by design, and one generally closed for national security reasons. For the most part, these differences can be overcome on S&T activities through appropriate collaborative planning.

Science and Technology Development

When our missions are common, when our technology requirements are similar, and when we can make the best use of our nation's limited space infrastructure, both NASA and the DOD benefit from cooperative efforts in S&T development. Whether maneuvering in space, experimenting in space or communicating in space, there is shared workspace that leads to the best possible equipment, processes and procedures to ensure success, whether the mission is military or civil in nature. In developing basic technologies for launch systems, materials for use in space activities, or developing infrastructure to command and communicate with our space assets, DOD and NASA cooperation is key to making the most of our space dollars.

When resources, missions, and technologies are purely military in nature, we are not best served by collaborating. It is important that our civil space activities be kept free from possible accusations of militarism. While many technologies developed by NASA–DOD collaboration are dual-use in nature, there are some developmental areas that should remain out of bounds.

NPOESS—One of the most important joint collaborative efforts currently underway between NASA and the Air Force is the National Polar-orbiting Operational Environment System (NPOESS) a tri-agency program of NASA, DOD, and the Department of Commerce (DOC) that converges the DOD and DOC/NOAA polar-orbiting weather satellite programs. NASA, working with NPOESS Integrated Program Office (IPO), is providing pre-operational risk-reduction demonstration and validation tests for four critical NPOESS sensors that will fly on the NPOESS Preparatory Project (NPP). NPP is a primary NASA mission that serves as a "bridge" between the Earth Observation Satellite (EOS) mission and NPOESS. NPP is also a critical risk reduction mission for the Visual Infrared Imager Radiometer Suite (VIIRS), the Cross-track Infrared Sounder (CrIS), the Advanced Technology Microwave Sounder (ATMS), and the Ozone Mapper/Profiler Suite (OMPS) sensors and serves as an end-to-end test for the Command, Control and Communication (C3) and data processing systems for NPOESS.

DOD Space Test Program—While DOD currently has no requirement for manned space flight, the S&T community, through the auspices of the DOD Space Test Program, has made excellent use of the Space Shuttle. When necessary, NASA and DOD have worked to develop new integration methods and hardware to make the most use of every ounce of available space lift. To date, the Space Test Program has launched over 200 experiments on over seventy Shuttle missions, including some of the first science experiments that were carried out on the international Space Station. As part of the effort to provide risk reduction to the NPOESS system, the DOD Space Test Program launched the Coriolis Mission in January 2003. This mission hosted both a solar mass ejection imager and WindSat sensor. The WindSat sensor will be evaluated for use on the NPOESS system.

NASA also assisted the Space Test Program in tests of a new Vibro-Acoustic Launch Protection Experiment (VALPE). NASA supported two successful sounding rocket launches from the Wallops Island launch facility. At Cape Canaveral, the AF supports the launch of NASA payloads, most recently the Spirit and Opportunity rovers now investigating Mars, from the Eastern and Western ranges.

S&T Forums—In more basic research and development, the AF and NASA collaborate in several major research projects and have several forums set up to facilitate S&T. The three major coordination forums for collaborative work are: the National Thermal Protection Systems (TPS) Working Group, which is led by the AF and NASA with participation of Army, Navy, Department of Energy (DOE), industry, and academia, and fosters development of new and advanced thermal protection materials and systems; the National Space and Missile Materials Symposium, which fosters increased communication and understanding in pursuing key materials technology challenges for space and missiles; and the Air Force Research Laboratory (AFRL)–Jet Propulsion Laboratory (JPL) Annual Summit, which is held annually to discuss and coordinate research efforts.

Other areas of research range across the complete spectrum of S&T activities. NASA and the AF work to track and characterize orbital debris as well as performing asteroid surveys to detect any large objects that are at risk of striking Earth. Many materials science experiments are carried out to look at environmental effects of space exposure, as in the DOD Space Test Program MISSE experiments, as well as high stress/high-cycle experiments on airframe or fuel tank materials. In the Integrated High Payoff Rocket Propulsion Technology (IHPRPT) Program, AFRL and NASA have worked to develop a spiral improvement system to the Space Shuttle Main Engines with technology benefits that will help all U.S. next generation rocket engines. In addition to all of these areas, NASA and the AF collaborate on other S&T programs that touch on almost every facet of both aviation and space technologies. For instance, AFRL is teaming with JPL to develop the L–Band antenna for NASA's space based radar effort.

Even NASA's Mars exploration mission benefits from AF collaboration with NASA utilizing AF-developed Rad-6000 32-bit microprocessors and lithium-ion batteries in both planetary rovers. In addition, AF operational studies provided expertise on human fatigue-related performance issues that will help provide counter-fatigue strategies for rover operators.

Space Operations

The AF and NASA have existing memoranda of agreement establishing partnerships to support NASA launches with Spacelift Range assets and to pursue advanced launch and test range technologies. The AF's Spacelift Ranges support all launch operations for NASA manned and unmanned launches from the Kennedy Space Center (KSC) or from Vandenberg AFB. Also, at the recommendation of the Interagency Working Group on Future Management and Use of the U.S. Space Launch Bases and Ranges, the AF and NASA established the Advanced Range Technology Working Group (ARTWG), co-chaired by AF Space Command and NASA–KSC. The ARTWG charter focuses on improving safety, increasing flexibility and capacity, and lowering range costs in support of future generations of reusable and expendable launch vehicles. The Joint Base Operating Support Contract (JBOSC) is a joint procurement effort between KSC NASA and the AF's 45th Space Wing (SW) to provide unified base support services for KSC, Cape Canaveral AFS, and Patrick AFB.

Human Space Flight

AFSPC provided support to NASA (via USSTRATCOM to USNORTHCOM) on the *Columbia* accident response and subsequent investigation. Major General John Barry and Brigadier General Duane Deal, USAF, both served on the *Columbia* Accident Investigation Board (CAIB). General Deal also heads the wing (21st SW) responsible for operating the space surveillance network, which assisted in the *Columbia* investigation. Approximately 20 AFRL personnel from six technology directorates participated in the CAIB via the DOD

Columbia Investigation Support Team. Subject matter expertise was provided in the fields of non-destructive inspection and test of critical composite structures, space weather, atmospheric space chemistry and physics, reentry physics, high-speed aerodynamics, aerothermal environments, kapton insulated wiring, ceramic materials, structural fatigue/fracture failure, and human behavior "group think" decision-making.

In an effort to assist NASA in its return-to-flight activities for the Shuttle fleet, the AF is assisting in developing and evaluating leading edge repair concepts that can be applied by astronauts in orbit. To date, 20 specimens from seven different organizations have been tested with three concepts surviving thermal conditions representative of flight heat flux and temperature. These three will be studied further to fully characterize the performance of the repair methods and materials and certify the concepts for flight. The AF is also assisting to analyze and improve the manual foam spraying operation previously used on the Space Shuttle *Columbia* external tanks.

NASA–DOD Space Organizations

Since 1958, the White House has created several organizational mechanisms to coordinate civil and military space programs and activities, including R&D investment. These range from President Eisenhower's Civilian-Military Liaison Committee, which was designed to coordinate NASA and DOD activities, to the Kennedy-Johnson-era National Space Council and Aeronautics and Astronautics Coordinating Board, to President Reagan's National Security Council-led interagency group, and then to President Clinton's decision to separate Space Council functions under the Office of Science and Technology Policy and National Security Council. The current organization mechanism for coordination between NASA and DOD, however, is the Partnership Council.

The U.S. can, and always will, explore better ways of coordinating NASA and DOD space activities. Analysts from both NASA and DOD routinely track developments in space management involving international partners in space cooperation as well as other spacefaring nations. While it is always beneficial to study how other countries attack similar problems, we must always be cognizant of the fact that other countries have different policies, laws, technologies and national security and civil requirements.

DSB/AFSAB Joint Task Force on Acquisition of National Security Space Programs

We in the DOD have benefited greatly from the recommendations of the joint Defense Science Board and Air Force Scientific Advisory Board task force on Acquisition of National Security Space Programs, led by Mr. A. Thomas Young. Mr. Young is a past Director of the Goddard Space Flight Center, and headed the 1999 NASA-chartered review of the Mars Polar Lander loss.

Just as in the DOD, during the 1990s, NASA experienced declining budgets, increased acceptance of risk (for example—Faster, Better, Cheaper), unrealized growth of a commercial space market, increased dependence on space by an expanding user base, and consolidation of the space industrial base.

The Young Panel identified five "basic reasons" for cost growth and schedule delays in National Security Space programs:

- Cost has replaced mission success as the primary driver in managing space development programs. . .resulting in excessive technical and schedule risk.
- Unrealistically low cost estimates lead to unrealistic budgets and unexecutable programs.
- Undisciplined definition and uncontrolled growth in system requirements.
- Government capabilities to lead and manage the acquisition process have seriously eroded.
- Industry has failed to implement proven practices on some programs.. . .The space industrial base is adequate to support current programs, although long-term concerns exist.

Within the DOD, we have taken the Young Panel findings and recommendations very seriously, and are continuing to implement policy and process changes in response to the Young Panel recommendations. Many of these findings likely have some applicability to NASA since we share much of the same industrial base and have experienced similar budget pressures. We have shared the Panel's results and our lessons learned with senior NASA leadership during the Partnership Council and our other interactions.

Conclusion

Historically, the DOD and NASA have fostered a collaborative relationship to maximize responsive access to space and national space investment strategies, and we will continue to do so in the future. Both organizations have benefited from this open exchange of ideas and lessons learned, laying the foundation for future collaborations.

NASA was formed with many DOD centers of excellence as its space-related core. The Mercury and Gemini missions, for instance, all flew on DOD launch vehicles. From that time forward, we have continued to collaborate across the full spectrum of space—launch, communications, sensors, materials, life sciences, and much more. In many respects, the relationships between NASA and the DOD are as close, or closer, than they have ever been.

Biography for Robert S. Dickman

Robert S. Dickman is Deputy for Military Space, Office of the Under Secretary of the Air Force, Washington, D.C. He supports the Under Secretary, who is also the Director of the National Reconnaissance Office, in executing space responsibilities, which include managing the planning, programming and acquisition of space systems for the Air Force and other military services.

Mr. Dickman was born in Brooklyn, N.Y., grew up in New Jersey, and entered the Air Force as a distinguished graduate of the ROTC program at Union College, Schenectady, N.Y. He has had a varied career in space operations, and acquisition and planning, including being assigned at the Space and Missile Systems Center, the Pentagon, North American Aerospace Defense Command, U.S. Space Command, Air Force Space Command, and the National Reconnaissance Office. While serving on active duty, he was the first Vice Commander of the 2nd (now 50th) Space Wing, Schriever Air Force Base, Colo., Commander of the 45th Space Wing, Patrick Air Force Base, Fla., Department of Defense Space Architect, and the senior military officer at the NRO in Washington, D.C. He retired from active duty in 2000 in the rank of major general, and was appointed to the Senior Executive Service in March 2002.

EDUCATION

1966—Bachelor's degree in physics, Union College, Schenectady, N.Y.

1968—Master's degree in space physics, Air Force Institute of Technology

1976—Distinguished graduate, Air Command and Staff College, Maxwell Air Force Base, Ala.

1978—National Security Management Course, National Defense University

1983—Master's degree in management, Salve Regina College, Newport, R.I.

1983—Distinguished graduate, Naval War College, Newport, R.I.

CAREER CHRONOLOGY

1. June 1966–June 1968, student, Air Force Institute of Technology
2. June 1968–June 1972, program manager for theoretical and particle physics, Air Force Office of Scientific Research, Arlington, Va.
3. July 1972–May 1973, satellite communications program element monitor, Directorate of Space, Headquarters U.S. Air Force, Washington, D.C.
4. June 1973–May 1975, terminal systems manager, Air Force Satellite Communications System Program Office, Los Angeles AFB, Calif.
5. June 1976–September 1979, operational manager for military satellite communications, Deputy Chief of Staff for Plans and Operations, Headquarters U.S. Air Force, Washington, D.C.
6. October 1979–January 1982, Chief, Implementation Branch, Space Defense Operations Center, Headquarters Aerospace Defense Command, Cheyenne Mountain AFB, Colo.
7. February 1982–June 1982, Executive Officer to the Vice Commander in Chief, North American Aerospace Defense Command, Colorado Springs, Colo.
8. July 1982–June 1983, student, Naval War College, Newport, R.I.
9. July 1983–June 1984, Director of Space Systems, Deputy Chief of Staff for Operations, Headquarters Air Force Space Command, Peterson AFB, Colo.
10. July 1984–June 1985, Chief, Commander's Group, Headquarters North American Aerospace Defense Command and Air Force Space Command, Peterson AFB, Colo.
11. July 1985–May 1986, Vice Commander, 2nd Space Wing, Falcon AFB, Colo.
12. June 1986–June 1987, Assistant to the Director of Operations, U.S. Space Command, later, Director of Missile Warning, Air Force Space Command, Peterson AFB, Colo.
13. July 1987–June 1989, Chief, Space Systems Division, Directorate of Space and Strategic Defense Initiative Programs, Washington, D.C.
14. July 1989–June 1992, Deputy Director of Space Programs, Office of the Assistant Secretary of the Air Force for Acquisition, Washington, D.C.
15. July 1992–June 1993, Director of Plans, Headquarters Air Force Space Command, Peterson AFB, Colo.
16. July 1993–January 1995, Commander, 45th Space Wing, and Director, Eastern Range, Patrick AFB, Fla., and Cape Canaveral Air Station, Fla.
17. February 1995–September 1995, Director of Space Programs, Office of the Assistant Secretary of the Air Force for Acquisition, Washington, D.C.
18. October 1995–June 1998, Department of Defense Space Architect, Washington, D.C.
19. June 1998–August 2000, Director, Office of Plans and Analysis, and System of Systems Architect; Director, Office of Architectures, Assessments and Acquisition; Director, Corporate Operations Office; and senior military officer, National Reconnaissance Office, Washington, D.C.
20. March 2002–present, Deputy for Military Space, Office of the Undersecretary of the Air Force, Washington, D.C.

BADGES

Master Space Badge

MAJOR AWARDS AND DECORATIONS

Defense Distinguished Service Medal
Distinguished Service Medal
Defense Superior Service Medal
Legion of Merit
Defense Meritorious Service Medal
Meritorious Service Medal with oak leaf cluster
Air Force Commendation Medal with oak leaf cluster

OTHER ACHIEVEMENTS

National Reconnaissance Office Gold Medal
1995 Ira Eaker Fellow, Air Force Association
1998 Astronautics Award, National Space Club

PROFESSIONAL MEMBERSHIPS AND ASSOCIATIONS

Air Force Association
U.S. Naval Institute
American Institute of Aeronautics and Astronautics

Chairman ROHRABACHER. Thank you very much. And we will have Ron Sega and, of course, Elon Musk, after this. I would—I am sorry I can't say short break. It is going to probably be around 40 minutes or 45 minutes. So I appreciate those of you who could stay around. Elon, you are out in California. I know you are enjoying yourself out there. And I will be out there tonight. But we will all be back in 45 minutes. This subcommittee is in recess.

[Recess.]

Chairman ROHRABACHER. All right. And so the hearing is called to order. Our next witness is the Honorable Ron Sega, Director, Defense Research and Engineering, also one of our beloved astronauts and someone who has seen this firsthand, which we, of course, appreciate your firsthand experience. And he is the Chief Technical Advisor to the Secretary of Defense for all scientific and technical matters, basic and applied research, and advanced technology development. You may proceed.

STATEMENT OF THE HONORABLE RONALD M. SEGA, DIRECTOR, DEFENSE RESEARCH AND ENGINEERING, DEPARTMENT OF DEFENSE

Dr. SEGA. Thank you, Mr. Chairman and Members of the Committee. I appreciate the opportunity to appear before you today to talk about the Department of Defense's programs in research and engineering, particularly in space and aeronautics and DOD's collaboration with the National Aeronautics and Space Administration. Thank you for allowing my written testimony to be submitted for the record.

In addition to discussing some of the specifics of the NASA–DOD collaboration, it is important to understand how the research and development program and our activities in space and aeronautics are integrated within the Department of Defense. We do have several mechanisms for coordination of the R&D activities between DOD and NASA. One that you heard about was Space Partnership Council, established in 1997. This Council meets regularly and coordinates space issues. And it is actually meeting more recently than it has in the past. As a member of the Space Partnership Council, I believe it is a productive forum to address the overarching DOD–NASA requirements and issues related to space.

One initiative from the Department of Defense and in collaboration with NASA is the National Aerospace Initiative. As I began in the fall of 2001, it came apparent that there are many studies and reports in progress, in draft state, and near completion, but there was a lack of integration among the various efforts. So it was our goal to look at integrating these activities. And we divided it into three areas: high-speed hypersonics, space access, and space technology. And through a series of workshops within government, all branches of the DOD services and agencies, and NASA looked at what technologies were a—currently—the state of technology, which ones were available and opportunities for the future, and then worked with industry and academia to establish the state of technology in these three areas and then to provide technology roadmaps.

One example of a crosscutting program is RASCAL. It is the Responsive Access Small Cargo Affordable Launch program from DARPA. It combines a high-speed, air-breathing first stage with a rocket-based upper stage and a small responsive satellite to demonstrate reusable, affordable, and responsive space access. So RASCAL is an innovative approach to space access.

We looked at continuing many programs, because they made sense, they were part of the roadmap ahead, one was the Integrated High Performance Turbine Engine Technology program, called IHPTET. It had been a sustained investment from DOD and NASA and industry since 1988, and it was building a technology base and meeting milestones. Another example of an ongoing program was a Hypersonic Flight Demonstration Program, called HyFly, a program that is funded by DARPA and the Navy, in collaboration with NASA, universities, and industry. I personally visited the Langley Research Center during the summer of 2002, and they were testing in the Mach 6.2 to Mach 6.5 range in the eight-foot tunnel and the effective altitudes of 85,000 to 100,000 feet. And I believe that the work was being done in a very positive and effective way.

Recent additional programs, after doing this analysis, were several. Two examples would include a Single Engine Demonstration program at DARPA, Air Force funded program in the same flight regime of Mach 7 with a first flight target around 2008. A second example is FALCON, which you have heard about earlier, that provides technologies toward small launch vehicles, a Common Aero Vehicle for thermal protection system and aerodynamics as well as a hypersonic cruise vehicle.

An ongoing program that looks at rocket propulsion is IHPRPT, Integrated High Payoff Rocket Propulsion Technology program. I concur with Bob Dickman's assessment that this is a good program. It is a 15-year effort focused on developing measurable, affordable, and goal-directed rocket propulsion technologies. We believe the payoffs could be quite positive in that program. An example inside of IHPRPT is an Integrated Powerhead Demonstration, IPD. This is a key demonstration that is a joint NASA–Air Force project, and it is scheduled for engine testing at the NASA Stennis Space Center in 2005. They have had four successful component demonstrations over the last 18 months. It is a new flow engine cycle, and it should enable an increase in rocket engine reliability and mission life as well as reducing maintenance time and cost.

In terms of developing a space science and technology strategy, the National Defense Authorization Act for fiscal year 2004 requires the Secretary of Defense to development, implement, and annually review and revise a space science and technology strategy. As the Director of Defense Research and Engineering, I am charged to jointly develop and implement this strategy with the Under Secretary of the Air Force, who is the Department of Defense's Executive Agent for space.

We are actively working with the Department's research laboratories, the Defense Advance Research Projects Agency, DARPA, National Reconnaissance Office, and the Missile Defense Agency through a space S&T strategy team to develop and implement this strategy.

And finally, the Department of Defense and NASA's research and development programs support building the technology base to enable future capabilities. Since the days of Chuck Yeager, the National Advisory Committee for Aeronautics, and the Bell X–1 that broke the sound barrier, the DOD has conducted a broad range of cooperative and collaborative programs with NACA and now NASA. Recently, the National Aerospace Initiative technology plans provided an integrated technology roadmap and outlined the requisite investments to enable critical military and civil capabilities. We are excited about the synergies that can be derived as we work collaboratively to achieve our common science and technology goals.

And thank you for allowing me to appear before your committee.

[The prepared statement of Dr. Sega follows:]

PREPARED STATEMENT OF RONALD M. SEGA

Introduction

Mr. Chairman, Members of the Committee, thank you for the opportunity to appear before you today to talk about the Department of Defense's (DOD) research and engineering programs in space and aeronautics and DOD's collaboration with National Aeronautics and Space Agency (NASA). In addition to discussing some specifics of DOD–NASA collaboration, it is also important to understand how the research and development (R&D) activities for space and aeronautics technologies within the Department of Defense are integrated. There are several mechanisms for coordination of R&D activities between DOD and NASA.

Space Partnership Council

Since 1997, the Space Partnership Council (SPC) has been, and continues to be, a very productive mechanism for DOD–NASA collaboration and program coordination. The SPC addresses overarching DOD–NASA requirements and issues related to space. The council is comprised of the following members:

Under Secretary of the Air Force/Director of National Reconnaissance Office
Commander of Air Force Space Command
Commander of United States Strategic Command
Director of Defense Research and Engineering
Administrator of NASA

The Council meets regularly and coordinates space issues, such as technology development to enable goals like transformational space access, and operational space capabilities.

National Aerospace Initiative

Collaborative efforts between DOD and NASA over the past several years have been encompassed in the National Aerospace Initiative (NAI). NAI is a focused effort to coordinate technology development and demonstrations in three key aerospace technology areas, which are the pillars of the NAI. The three pillars are high speed and hypersonic flight; space access; and space technologies. Beginning as a concept in 2001, NAI has matured and supported development of integrated technology plans. One program that highlights the potential synergy gained between the pillars is the Responsive Access, Small Cargo, Affordable Launch (RASCAL) DARPA program. RASCAL is a program that combines a high speed air breathing first stage, with rocket-based upper stages, and a small responsive satellite to demonstrate a reusable, affordable, responsive space access system. RASCAL is a five year program to demonstrate the feasibility of coupled high speed/hypersonic flight, affordable access to space and small payload systems. Beyond RASCAL, extensive collaborations have occurred in research and development in all three areas. Through a series of workshops convened by DOD and NASA, which were followed by input from outside the government, detailed goals, objectives, technical challenges and approaches were developed.

NAI supports many important continuing programs such as the Integrated High Performance Turbine Engine Technology (IHPTET)/Versatile Affordable Advanced Turbine Engines (VAATE) projects. VAATE is a successful collaborative program that started in 1988, and involved DOD, NASA, and industry to have a long-term, focused research program to improve turbine engine technology. IHPTET is currently developing a common core to be used in the various commercial and military engines. The industry match has been an important component of the in IHPTET program. This turbine engine technology development is essential to many future government and commercial aerospace systems.

Each of the three pillars has significant activity. For high speed/hypersonic flight, the Hypersonic Flight Demonstration Program, known as HyFly, is a jointly funded program by DARPA and Office of Naval Research. The objective of HyFly is to develop and demonstrate, in flight, advanced technologies for hypersonic flight with near-term emphasis on a missile application. The HyFly hypersonic strike missile demonstrator vehicle is powered by a Dual Combustion Ramjet (DCR) engine. A DCR engine performance at Mach 6.5 was demonstrated on a full-scale model in freejet testing at NASA Langley Research Center in 2002. Its first powered flight in the atmosphere is expected in approximately one year.

Another example of an advanced prototype hypersonic missile is the Single Engine Demonstration (SED). SED will integrate the United States Air Force Hypersonic Technology (HyTech) engine with air vehicle technologies developed by Defense Advanced Research Projects Agency. The project involves government, industry, and academic hypersonic researchers and builds on previous DOD–NASA efforts. This exciting new demonstration will be flight tested by the end of the decade. The flight vehicle will be propelled by a hydrocarbon supersonic combustion ramjet (scramjet), and should ultimately achieve a Mach 7 to 8 flight. Success of HyFly and SED could enable a new aviation flight regime, historically analogous to the revolutionary introduction of the jet engine to propeller-driven aircraft.

The second area of significant collaboration is in our access to space access pillar. A long-term government/industry effort for advancing rocket propulsion is the Integrated High Payoff Rocket Propulsion Technology (IHPRPT) program. The IHPRPT is a three phase, 15-year national program to double space/missile propulsion capability, decrease cost and increase reliability by 2010, using government-industry partnership. A key element under IHPRPT is the joint NASA–Air Force project called the Integrated Powerhead Demonstration (IPD), which should culminate in the completion of engine testing at NASA's Stennis Space Center in 2005. Four successful component demonstrations have occurred in the past 18 months. This new liquid engine cycle should enable a 25 percent increase in rocket engine reliability, a 200-mission life for the engine, and a reduction in maintenance time and cost. The DOD–NASA cooperation, leading to the IPD full-flow cycle engine, should result in enhanced reusable and expendable space vehicle propulsion.

Another program which is jointly funded by DARPA and the Air Force is known as FALCON (Force Application and Launch from CONUS). FALCON is a new program to develop a Small Launch Vehicle (SLV), a Common Aero Vehicle (CAV), and a Hypersonic Cruise Vehicle (HCV). An initial goal is a rocket boosted glide vehicle capable of delivering 1,000 pounds at a distance of 3,000. Initial phases of FALCON are on-going and will demonstrate the aerodynamic properties of the flight vehicles. This program is envisioned to mature to a hypersonic glide plane capable of delivering 12,000 lbs. over 9,000 miles. Thus, the FALCON program should demonstrate and validate in-flight technologies that should enable both a near-term and far-term capability to execute time-critical, prompt global reach missions while at the same time, demonstrating affordable and responsive space lift.

Space Science and Technology Strategy

The National Defense Authorization Act for Fiscal Year 2004 requires that the Secretary of Defense develop, implement and, annually review and revise a space science and technology (S&T) strategy. As the Director of Defense Research and Engineering, I am charged to jointly develop and implement this strategy with the Under Secretary of the Air Force, who is the Department of Defense's Executive Agent for Space. The space S&T strategy is focused on short-term and long-term goals within the Department, the process of achieving these goals, and the process for assessing these goals. We are actively working with the Department's research laboratories and the Defense Advanced Research Projects Agency (DARPA), National Reconnaissance Office (NRO) and Missile Defense Agency (MDA) through a space S&T strategy team to develop and implement this strategy This Space Science and Technology Strategy will be incorporated in the National Security Space Plan.

Conclusion

The Department of Defense and NASA research and development programs support building the technology base to enable future capabilities. Since the days of Chuck Yeager and the National Advisory Committee for Aeronautics (NACA) X–1 that broke the sound barrier, the DOD has conducted a broad range of cooperative and collaborative programs with NACA now known as NASA. Recently, the National Aerospace Initiative technology plans have provided integrated technology roadmaps, and outlined the requisite investments to enable critical military and civil capabilities. We are excited about the synergies that can be derived as we work collaboratively to achieve our common science and technology goals and transformational objectives.

BIOGRAPHY FOR RONALD M. SEGA

The Honorable Ronald M. Sega, Director of Defense Research and Engineering (DDR&E), is the chief technical advisor to the Secretary of Defense and the Under Secretary of Defense for Acquisition, Technology, and Logistics (USD–AT&L) for scientific and technical matters, basic and applied research, and advanced technology development. Dr. Sega also has management oversight for the Defense Advanced Research Projects Agency (DARPA). [Defense Research and Development official functions]

Dr. Sega has had an extensive career in academia, research, and government service. He began his academic career as a faculty member in the Department of Physics at the U.S. Air Force Academy. His research activities in electromagnetic fields led to a Ph.D. in Electrical Engineering from the University of Colorado. He was appointed as Assistant Professor in the Department of Electrical and Computer Engineering at the University of Colorado at Colorado Springs in 1982. In addition to teaching and research activities, he also served as the Technical Director of the Laser and Aerospace Mechanics Directorate at the F.J. Seiler Research Laboratory and at the University of Houston as the Assistant Director of Flight Programs and Program Manager for the Wake Shield Facility. Dr. Sega became the Dean College of Engineering and Applied Science, University of Colorado at Colorado Springs in 1996. Dr. Sega has authored or co-authored over 100 technical publications and was promoted to Professor in 1990. He is also a Fellow of the Institute of Electrical and Electronic Engineers and the Institute for the Advancement of Engineering.

In 1990, Dr. Sega joined NASA, becoming an astronaut in July 1991. He served as a mission specialist on two Space Shuttle Flights, STS–60 in 1994, the first joint U.S. Russian Space Shuttle Mission and the first flight of the Wake Shield Facility, and STS–76 in 1996, the third docking mission to the Russian space station Mir where he was the Payload Commander. He was also the Co-Principal Investigator for the Wake Shield Facility and the Director of Operations for NASA activities at the Gagarin Cosmonaut Training Center, Russia, in 1994–95.

Dr. Sega has also been active in the Air Force Reserves. A Command Pilot in the Air Force with over 4,000 hours, he has served in various operational flying assignments, including a tour of duty as an Instructor Pilot. From 1984 to 2001, as a reservist assigned to Air Force Space Command (AFSPC), he held positions in planning analysis and operational activities, including Mission Ready Crew Commander for satellite operations—Global Positioning System (GPS)—Defense Support Program (DSP), and Midcourse Space Experiment (MSX), etc. He was promoted to the rank of Major General in the Air Force Reserves in July 2001.

Chairman ROHRABACHER. Thank you very much. And I have some questions for you when we get back—or get past our final witness, which is—who is, I might say, Elon Musk, the CEO of SpaceX, an entrepreneurial launch company developing a new family of rockets. He is speaking to us on a video link from Los Angeles Air Force Base. And we welcome you, Mr. Musk, and you may proceed.

STATEMENT OF MR. ELON MUSK, CHIEF EXECUTIVE OFFICER, SPACE EXPLORATION TECHNOLOGIES

Mr. MUSK. Thank you, Mr. Chairman. I hope you can hear me okay.

It is an honor to be here with Admiral Steidle, Bob Dickman, Dr. Sega, and Ms. Poniatowski, I guess in the question period.

It is common knowledge that the U.S. launch industry is non-competitive. An appropriate comparison is the U.S. auto industry of the 1970s, prior to entry of the Japanese. However, that would be quite generous. At no point during that period did General Motors decide, as Boeing has recently done, that they would only service government customers.

In the case of launch vehicles, the noncompetitiveness is so great that SpaceX is confident of not just a significant improvement in reliability, but also of maintaining a several fold price reduction. Hopefully, this will stimulate the other three U.S. launch vehicle companies to reexamine their processes, as GM and Ford did in their time, and provide a better and lower cost product to their customers.

I am also optimistic that the success of SpaceX will result in other entrepreneurial companies entering the space business, both in launch and the manufacture of lower cost spacecraft. Some look at the cost of launch and comment that it only represents a portion of the total mission cost. This is a very naïve conclusion. In fact, it all starts with launch. If you are paying $5,000 a pound for something in orbit, you will naturally pay up to $5,000 a pound to save weight on your satellite, creating a vicious circle of cost inflation.

The result is a cost impedance match between the spacecraft and the launch vehicle, but it is driven by the launch vehicle. If you could launch for much lower cost and manifest quickly, that satellite would cost a lot less. A case example is TacSat–1, the DOD satellite on the maiden flight of Falcon I.

The benefits and risks for the U.S. launch industry of NASA–DOD collaboration, the most significant would be automatic cross-certification of a new launch vehicle. If a launch vehicle is found to be satisfactory for a DOD satellite, then it should be satisfactory for NASA, and vice versa.

That is currently not the case. For example, the Boeing Delta IV and Lockheed Atlas V have had to undergo separate DOD and NASA on-ramp processes. The result is greater expense to the taxpayer and those companies. SpaceX is in a similar position where we are undergoing a DOD review of our Falcon launch vehicle by the Aerospace Corporation, but will later have to repeat the process for NASA

The biggest risk to launch vehicle development from NASA–DOD collaboration would be excessive requirements accumulation, as occurred with the Space Shuttle. In my experience, having personally developed extremely complex technology systems, it is critical that the number of people determining requirements be kept very small and consist of only the most talented and experienced personnel. Otherwise, one may be faced with requirements that are easily addressed individually, but not combined. Asking that a product serve as either a floor wax or a dessert topping is fine, but not both at the same time. It won't be a tasty dish.

Recommendations for how NASA and the DOD can support emerging U.S. launch companies. Okay. Number one, buy an early launch. Although our launch vehicle has been an entirely private development, the DOD has been very supportive by purchasing the first flight. Here we would like to express our thanks to Air Force Space Command, the Force Transformation group in the Office of the Secretary of Defense, and the Naval Research Laboratory. Our country should be proud of those organizations and what they are doing to strengthen our capabilities in space.

To date, there has been limited dialogue with SpaceX initiated by the launch vehicle procurement office of NASA. This may be a function, historically, of the Code M office believing that they have no mandate to foster new U.S. launch providers. We recommend that this be established as an explicit goal and that NASA offer to buy the first, or at least an early launch, of a new vehicle. A promising sign is the funding allocation in the proposed NASA budget, referred to as the Small Payload Launch Initiative.

All right. Number two, streamline the regulatory process. Obtaining approval to launch from the government ranges is a very complex process. Once SpaceX has completed this process for the Falcon, we will work with the Air Force to provide recommendations for how it can be streamlined for other emerging launch providers. This will not benefit SpaceX, as we already will have had our approval. We will do it simply because it is a good thing for our country, and the cause of space exploration will be greater served.

Number three, increase and extend the use of prizes. The strategy of offering prizes for achievements in space technology can pay enormous dividends. History is replete with examples of prizes spurring great achievements, such as the Orteig Prize, famously won by Charles Lindbergh, and the longitude prize for ocean navigation. The subjectivity and error of proposal evaluation is removed, and the solution may be, in a way and from a company, that no one ever expected. We strongly endorse and urge Congress to support and extend the proposed Centennial Prizes put forward in the recent NASA budget. No dollar spent on space research will yield a greater value for the American people than those prizes.

What unique capabilities might emerging launch providers offer to NASA and the DOD? Well, number one, reliability. Current launch vehicles are considered to be "reliable" if their failure rate is only one in fifty. In any other mode of transport, this would be considered outrageously unreliable. New companies might ultimately provide reliability levels more comparable with airline transportation.

In the case of SpaceX, we believe that our second-generation vehicle, in particular, the Falcon V, will provide a factor of ten improvement in propulsion reliability. Falcon V will be the first U.S. launch vehicle since the Saturn V Moon rocket that can complete its mission even if an engine fails in flight, like almost all commercial aircraft.

Number two, cost. Citing an inability to sell rockets commercially, the incumbent launch providers are dramatically increasing their prices, forcing NASA and the DOD to do fewer missions while paying more and more. The effect is material, severe, and gets worse every year. For a given budget, this obviously results in being forced to cancel missions that might otherwise have flown. Apart from public relations, there is no practical difference between a mission that was canceled for cost reasons and one that failed for other reasons. Either way, you have lost the mission.

In contrast, the SpaceX launch vehicles are commercially competitive worldwide in price and are only a fraction of the cost of our U.S. competitors. Moreover, we expect to decrease our prices in real, if not absolute terms every year.

Factoring in overhead, as anyone could tell by visiting our headquarters, SpaceX can provide a launch vehicle at half the price of both Boeing or Lockheed. We have made significant strides in each of the technical cost drivers, which I would be happy to address in the question period.

Number three, responsiveness. If space assets are needed to cover a particular geography or replace an unexpected loss of coverage, they can not be deployed in time with the existing launch providers. Emerging launch vehicle companies can provide, and dramatically improve, response time.

Four, flight environment. Existing rockets provide a terrible flight environment for satellites that is extreme in noise, vibration, shock, and g loading. These factors drive much of a satellite's design, despite the fact that it sees these loads for only the first 10 to 15 minutes required to reach orbit. For the remaining years of life, being in microgravity, the satellite sees essentially zero load.

New launch vehicles, like the Falcon, provide a much better flight environment, thus making the satellite design easier and the satellite itself more likely to reach orbit.

Thank you.

[The prepared statement of Mr. Musk follows:]

PREPARED STATEMENT OF ELON MUSK

How would you characterize the state of the U.S. space launch industry?

It is common knowledge that the U.S. space launch industry is fundamentally uncompetitive. An appropriate comparison one could draw is the U.S. auto industry of the 1970's, prior to entry of the Japanese. However, that would be quite flattering. At no point during that period did General Motors decide, as Boeing has recently done, that they would only service government customers.

One must be cautious, therefore, in reaching launch vehicle economics conclusions that are based on historical U.S. costs. What the reliability and price of launch should be cannot be determined by looking at Boeing and Lockheed, any more than one could properly draw conclusions about automobile reliability and pricing by looking at a 1975 Pinto or Cadillac.

Please note that I emphasize and place reliability ahead of price. The Japanese automobiles, especially in the 1980's with the adoption of total quality management techniques, were not just lower cost, but also of much greater reliability. The latter was arguably a bigger determinant of their success than price.

In the case of launch vehicles, the level of uncompetitiveness is so great that we at SpaceX are confident of not just a significant improvement in reliability, but also of establishing and maintaining a several fold price reduction. Hopefully, this will stimulate the other three U.S. launch vehicle companies to re-examine their processes, as GM and Ford did in their time, and provide a better and lower cost product to their customers.

I am also optimistic that the success of SpaceX will result in other entrepreneurial companies entering the space business, both in launch and the manufacture of lower cost spacecraft. Some look at the cost of launch and comment that it only represents a portion of the total mission cost. This is a very naïve conclusion. In fact, it all starts with launch cost. If you are paying $5000/lb. to put something in orbit, you will naturally pay up to $5000/lb. to save weight on your satellite, creating a vicious circle of cost inflation.

The result is a cost impedance match between the spacecraft and the launch vehicle, but it is driven by the launch vehicle. If you could launch for much lower cost and manifest quickly, instead of the two years advance notice required to launch in the U.S., that satellite would cost a lot less. A case example is TacSat–1, the DOD satellite on the maiden flight of Falcon I.

What are the benefits and risks for the U.S. domestic launch industry, including emerging U.S. launch vehicle providers, if NASA and the Department of Defense (DOD) collaborated more in the development and purchases of launch vehicles?

The most significant benefit for the U.S. launch industry from greater NASA–DOD collaboration would be automatic cross-certification of a new launch vehicle. If a launch vehicle is found to be satisfactory for launch of a Department of Defense satellite, then it should be satisfactory for NASA and vice versa.

That is currently not the case. For example, in the EELV program, both the Boeing Delta IV and Lockheed Atlas V have had to undergo separate DOD and NASA certification or on-ramp processes. The result is greater expense to the taxpayer and the aforementioned companies. SpaceX is in a similar position, where we are undergoing a DOD review of our Falcon launch vehicle by Aerospace Corporation, but will later have to repeat the process for NASA.

The biggest risk to a launch vehicle development from NASA–DOD collaboration in a development program would be excessive requirements accumulation, as occurred with the Space Shuttle. In my experience, having personally developed extremely complex technology systems, it is critical that the number of people determining requirements be kept very small and consist of only the most talented and experienced personnel. Otherwise, one may be faced with requirements that are easily addressed individually, but not combined. Asking that a product serve as either floor wax or a dessert topping is fine, but not both at the same time.

What specific recommendations would you make for how NASA and the DOD can encourage the healthy growth of the U.S. domestic launch market, especially for emerging commercial launch providers?

Buy an Early Launch

Although our Falcon launch vehicle has been an entirely private development, the DOD has been very supportive by purchasing the first flight. Here we would like to express our thanks to Air Force Space Command, the Force Transformation group in the Office of the Secretary of Defense and the Naval Research Laboratory. Our country should be proud of those organizations and what they are doing to strengthen our capabilities in space.

To date, there has been limited dialogue with SpaceX initiated by the launch vehicle procurement office of NASA. This may be a function historically of the Code M office operating under the assumption that they have no mandate to foster new U.S. launch providers. We recommend that this be established as an explicit goal and that NASA offer to buy the first or at least an early launch of a new vehicle, even if only on a success contingency basis. A promising sign is the funding allocation in the proposed NASA budget referred to as the Small Payload Launch Initiative.

Streamline the Regulatory Process

Obtaining approval to launch from the government ranges is a very complex and arduous process. Once SpaceX has completed this process for the Falcon I, we will work with the Air Force to provide a series of recommendations for how this can be streamlined, without sacrificing safety, for other emerging launch providers. Please note that this will not benefit SpaceX, as we will have already received our approval. We will do it simply because it is a good thing for our country and the cause of space exploration will be greater served.

Increase and Extend the Use of Prizes

The strategy of offering prizes for achievements in space technology or launch vehicle development milestones can pay enormous dividends. We are beginning to see how powerful this can be by observing the recent DARPA Grand Challenge and the X–Prize. History is replete with examples of prizes spurring great achievements, such as the Orteig Prize, famously won by Charles Lindbergh, and the Longitude prize for ocean navigation.

Few things stoke the fires of American creativity and ingenuity more than competing for a prize in fair and open competition. The result is an efficient Darwinian exercise with the subjectivity and error of proposal evaluation removed. The best means of solving the problem will be found and that solution may be in a way and from a company that no-one ever expected.

We strongly endorse and urge Congress to support and extend the proposed Centennial Prizes put forward in the recent NASA budget. No dollar spent on space research will yield greater value for the American people than those prizes.

What unique capabilities do emerging launch vehicle providers, like SpaceX, provide to NASA and the DOD?

The service of space transportation is defined by four variables: reliability, cost, responsiveness and payload environment. Emerging launch vehicle providers can provide breakthroughs in all areas.

Reliability

Current launch vehicles are considered by NASA and the DOD to be "reliable" if their failure rate is only one in fifty. In any other mode of transport, this would be considered outrageously *un*reliable. New companies might ultimately provide reliability levels more comparable with airline transportation.

In the case of SpaceX, we believe that our second generation vehicle in particular, the Falcon V, will provide a factor of ten improvement in propulsion reliability. Falcon V will be the first U.S. launch vehicle since the Saturn V Moon rocket that can complete its mission even if an engine fails in flight—like almost all commercial aircraft. In fact, Saturn V, which had a flawless flight record, was able to complete its mission on two occasions only because it had engine out redundancy.

Cost

Citing an inability to sell rockets commercially, the incumbent launch vehicle providers are dramatically increasing their prices, forcing NASA and the DOD to do fewer and fewer missions while paying more and more. The effect is material, severe and gets worse every year. For a given budget, this obviously results in being forced to cancel missions that might otherwise have flown. Apart from public relations, there is no practical difference between a mission that was canceled for cost reasons and one that failed for other reasons. Either way, you have lost the mission.

In contrast, the SpaceX launch vehicles are commercially competitive worldwide in price and are only a fraction the cost of our U.S. competitors. Moreover, we expect to decrease our prices in real, if not absolute, terms every year.

Launch vehicle pricing is driven by five factors: company overhead, engine costs, airframe costs, avionics costs and launch operations (including payload integration and range costs). Factoring in overhead alone, as anyone could tell by visiting our headquarters, SpaceX can produce a launch vehicle at half the price of Boeing or Lockheed. We have also made significant strides in each of the technical cost drivers, although time does not allow me to address each in detail. I would be happy to do so in the question period.

Responsiveness

The minimum time from contract signing to launch for incumbent U.S. launch companies is approximately two years. For the DOD in particular, this means a very constrained ability to respond quickly to threats as they develop. If space assets are needed either to cover a particular geography or replace an unexpected loss of coverage, they cannot be deployed in time. Emerging launch vehicle companies, like SpaceX, will provide a response time measured in months or weeks.

Payload Flight Environment

Existing rockets provide a terrible flight environment for satellites that is extreme in noise, vibration, shock and g-loading. These factors drive much of a satellite's design, despite the fact that it only sees these loads for the 10 to 15 minutes required to reach orbit. For the remaining years of life, being in microgravity, the satellite sees essentially zero load.

New launch vehicles, like the Falcon, provide a much better flight environment, thus making the satellite design easier and the satellite itself more likely to reach orbit safely.

Biography for Elon Musk

SpaceX is the third company founded by Mr. Musk. Prior to SpaceX, he co-founded PayPal, the world's leading electronic payment system, and served as the company's Chairman and CEO. PayPal has over twenty million customers in 38 countries, processes several billion dollars per year and went public on the NASDAQ under PYPL in early 2002. Mr. Musk was the largest shareholder of PayPal until the company was acquired by eBay for $1.5 billion in October 2002.

Before PayPal, Mr. Musk co-founded Zip2 Corporation in 1995, a leading provider of enterprise software and services to the media industry, with investments from The New York Times Company, Knight-Ridder, MDV, Softbank and the Hearst Corporation. He served as Chairman, CEO and Chief Technology Officer and in March 1999 sold Zip2 to Compaq for $307 million in an all cash transaction.

Mr. Musk's early experience extends across a spectrum of advanced technology industries, from high energy density ultra-capacitors at Pinnacle Research to software development at Rocket Science and Microsoft. He has a physics degree from the University of Pennsylvania, a business degree from Wharton and originally came out to California to pursue graduate studies in energy physics at Stanford.

DISCUSSION

NASA'S POLICIES TOWARD THE USE OF NEW LAUNCH VEHICLES

Chairman ROHRABACHER. Thank you very much, Mr. Musk.

And we will have some questions now. I would like to ask—first of all, thank you to all of the witnesses for their testimony, of course, but before we proceed, I would like to ask Ms. Karen Poniatowski to come to the witness table. And as NASA's Assistant Associate Administrator for Launch Services, she will be joining the Admiral in answering some of these questions. And of course, the first question, which I will pose to both of you, is NASA's. What we have heard today, especially from Mr. Musk just there, is that current NASA policy forbids NASA to contract for launch services unless the type of rocket being used has performed at least one successful flight. NASA's policy was put in place in the mid-1990s after several rockets failed.

Now we have heard that the DOD does not have this same policy. And in fact, Mr. Musk, who has spent a considerable amount of his own money, has been investing in a new rocket system, because he is being given the opportunity by the Department of Defense. Now shouldn't NASA be providing this same sort of incentive for people like Mr. Musk to invest their money into launch systems? And you may answer those questions.

Rear Admiral STEIDLE. Yes, sir. Thank you. Yes, you are accurate. That is the policy. What we have launched is generally one-of-a-kind, significant investment, and Karen was here when that policy was put together, so she is going to extrapolate from here on where and why we are doing that.

Ms. PONIATOWSKI. Yeah. Thank you for the opportunity, Mr. Rohrabacher and Committee, to sort of address some of these issues, particularly with the emerging companies. I followed the small launch companies for, I hate to say it, almost 20 years now. I started when I was very, very young.

Mr. MUSK. I am having a problem. It is very difficult to hear unless you speak directly to the mike.

Ms. PONIATOWSKI. Okay. Is that better?

Mr. MUSK. Yes, that is much better. Thank you.

Chairman ROHRABACHER. All right. Go right ahead.

Ms. PONIATOWSKI. Okay. And so what we have been watching is an ebb and flow in this particular market class. If you go back to the early '90s, you will see, with DARPA, they developed an emerging launch capability that was Orbital Science's Pegasus, and it had some initial start-up failures but then became a very robust, reliable system. In the mid-'90s, NASA sponsored and did fly on the first and only launch of the Conestoga commercial launch vehicle. Unfortunately, that ended in failure. We flew on the first of the Athena I launch vehicles after their first test flight was a failure. We flew the first Athena II mission, and so we do, indeed, have a history of flying on vehicles with no flight history.

Chairman ROHRABACHER. That—does that cost NASA a lot of money to do that, is that why the policy changed?

Ms. PONIATOWSKI. No, not at all. What the policy actually did is it tried to say, "We need to take flight history into consideration." The actual policy allows us to fly payloads with no—to fly on vehicles with no flight history. What has happened—what we did is we set up a process that identified payloads that had the level of risk that could tolerate a first flight with a new vehicle, missions that needed at least one flight, and those high-value kinds of missions that needed a more demonstrated flight history as one of the conditions when we looked at making a mission assignment.

What has happened is, over the past few years, we don't have many requirements that have been able to tolerate that risk, and in tandem with that, as we have worked with the emerging community of which, at any given time, there is nine to ten different entities that would like to enter the rocket business, we have seen an influx of international capability in the small class that has really hurt the domestic capability in that market niche. And that really has been the biggest threat that we have seen for some of the emerging companies not being able to get access. There is not a lot of demand. On average, NASA's requirements in this class run one to two flights per year. The DOD is in a similar position, so the overall demand is not particularly robust in this kind of a class. But we do look forward to working with vehicles, companies like SpaceX. We have met with Elon a number of occasions.

The other is NASA is—will not be able to be the first launch on Falcon since the DOD got there first, but we do have a small technology payload, Spacetech VI, which had been planning to fly as an instrument on a commercial bus team encounter, they had been looking at making arrangements for a launch, and they are subsequently now, I believe, in discussions with Mr. Musk on flying that——

Chairman ROHRABACHER. Okay. Well, let me get this straight in what we are saying here. You are suggesting that there has not been a policy of no first use, but that you judge—you are making judgments based on risk——

Ms. PONIATOWSKI. Right.

Chairman ROHRABACHER.—and you just haven't found one yet, or have you found one that I don't know about, that is worth the risk?

Ms. PONIATOWSKI. And as I said, the Spacetech VI, the current payload we are discussing right now——

Chairman ROHRABACHER. Correct.

Ms. PONIATOWSKI. Correct.

Chairman ROHRABACHER. Correct. But over the last, what, five or 10 years, that has not been the case.

Ms. PONIATOWSKI. That is correct.

Chairman ROHRABACHER. Okay. So this is the first one in how many years?

Ms. PONIATOWSKI. 1997.

Chairman ROHRABACHER. All right. So about six or seven years now.

Ms. PONIATOWSKI. Yeah, we did have—I think you are familiar with the University Explorer program. And those were small payloads we were trying to look at in that capability. It ended up, at that point, there weren't any small launch vehicles, so we flew those as half of the Pegasus——

Chairman ROHRABACHER. How many satellites do we have waiting to be launched?

Ms. PONIATOWSKI. In this payload class, on average, about one to two per year.

Chairman ROHRABACHER. Okay. How many do we have waiting now to be launched?

Ms. PONIATOWSKI. I don't have any that aren't tied to a mission vehicle. I have none pending.

Chairman ROHRABACHER. Okay. There are no pending satellites of this class that need a transportation system?

Ms. PONIATOWSKI. Not that aren't assigned right now.

Chairman ROHRABACHER. Okay. Because all of these other satellites that are waiting are larger satellites that are——

Ms. PONIATOWSKI. Right. Correct. Yeah, we have got some that we are starting to look at new missions downstream, but right now, the missions we have are manifested on a vehicle today.

Chairman ROHRABACHER. Okay. Mr. Musk, does that satisfy you and the private sector, that answer?

Mr. MUSK. If it were the—that the TMA counter satellites, I think we are working through that possibility of launch, and hopefully that works out. The sense I get though, and we can provide more detail on this outside of this forum, is that there are actually—that there are more satellites that wish to go up than one or two a year. And in particular, there were satellites and payloads and so forth that might otherwise have gone up on the Space Shuttle, which obviously could not go up on the Space Shuttle today, and even when it is flying, there is quite a backlog. So I——

Chairman ROHRABACHER. So there is a backlog? Now you are suggesting, is there a backlog of the smaller satellites that—is that what you are saying?

Mr. MUSK. I—that is my understanding, yes.

Chairman ROHRABACHER. We have got a little contradiction here. How about that, Karen?

Mr. MUSK. That is my idea.

Ms. PONIATOWSKI. Yeah. There are instruments——

Mr. MUSK. The biggest distinction between satellites which are destined to go up by themselves and satellites which might otherwise have gone up as hitchhikers or as payloads on the Space Shuttle, which still need to go up, but are not considered distinct travels by themselves.

Chairman ROHRABACHER. All right. Let me see——

Ms. PONIATOWSKI. Yeah.

Chairman ROHRABACHER. Let me get to the heart of that matter. Were you just referring to all satellites or are you referring to just the ones that aren't going to be hitchhiking? I mean, are there satellites that are small satellites that were going up on the Shuttle that you weren't counting?

Ms. PONIATOWSKI. Yeah. The difference is when you fly an expendable launch vehicle, it means that you bring with you a bus that has your power, your resources to be able to deploy that spacecraft and then use it in space. Many—most of the payloads that we flew as secondaries on the Shuttle, they take their power from the Shuttle, and so you can't transpose something like the gascan payloads that we fly, those are on a one for one transfer to be able to say you are going to fly on an expendable space vehicle.

Chairman ROHRABACHER. So these satellites that Mr. Musk is referring to——

Ms. PONIATOWSKI. Right.

Chairman ROHRABACHER.—are not satellites that a rocket that he is developing could put into space?

Ms. PONIATOWSKI. Not without an additional reformatting of those payloads. Some of them need to be returned. Some of those payloads need to have intervention; they need a switch turned on or off. And these types of payloads have tended to be very small, 50-kilogram types of payloads.

Chairman ROHRABACHER. Um-hum.

Ms. PONIATOWSKI. In the case of gas, you have got predominately university community in that the payload value—what they do is they basically pay us $50,000 for that kind of a——

Chairman ROHRABACHER. And that has to be all done on the Shuttle you are suggesting?

Ms. PONIATOWSKI. Correct.

Chairman ROHRABACHER. Is that your understanding, Mr. Musk, that all of these things have to be done on the Shuttle?

Mr. MUSK. Well, they have basically been intended for the Shuttle, and in some cases you get satellites which are relatively self-contained, and in some cases, they are more sort of instruments, but there is still that need that exists for them to go to space. And the Shuttle is not going to be able to meet that need, which I think, therefore, points one in the direction of adding the necessary functionality to those payloads—to those satellites as such that they can be launched on something like the Falcon I or other launch vehicles, because the alternative is that nothing happens to them and they stay on Earth.

Chairman ROHRABACHER. I think—well, why don't we give Karen the last word for this one?

Ms. PONIATOWSKI. Yeah, and that is why what the agencies put forward is the new Payload Launch Initiative, which is allowing us to put some money in the budget to start looking at some of these new flight opportunities that may arise, again, and partnering with both DARPA and with the Air Force Space Test Program, ways that we might be able to join some of these and do some joint missions and fly some of those payloads. That is why, as we have been retiring the capability on the Shuttle, we are looking at can we prime the pump, so to speak, and make some opportunities for some of these missions.

Chairman ROHRABACHER. I would like to end this part of the discussion, and then I will go to Mr. Lampson, with a general. It seems like the Air Force is willing to do this, but we are not willing—we are not getting much response from the Navy and NASA here. What——

Major General DICKMAN. Well, if the Navy doesn't launch their own——

Chairman ROHRABACHER. I am sorry. I shouldn't blame the Admiral for the Navy. I should just—you are in NASA now, Admiral.

Major General DICKMAN. The unwritten policy of the DOD is not particularly different than the one NASA uses. We judge the material of the launch vehicle and the mission criticality of the payload. For example, the first heavy-lift EELV, Delta IV, will fly without a payload, because we are going to do a demonstration flight. We weren't willing to fly that vehicle with a payload on it. So we make the same assessment that Karen and her team often does, and then chose to fly a demonstration.

Chairman ROHRABACHER. But for a smaller payload, you have been willing to be a lot more——

Major General DICKMAN. That is correct. It is——

Chairman ROHRABACHER.—courageous.

Major General DICKMAN. It is a great demonstration of both a launch vehicle and a responsive spacecraft that is worth the risk.

Chairman ROHRABACHER. All right. Mr. Lampson. And there will be a second round of questions.

IMPACTS OF THE PRESIDENT'S SPACE EXPLORATION INITIATIVE ON NASA'S SPACE LAUNCH INITIATIVE

Mr. LAMPSON. Admiral Steidle, one of the results of the President's space initiative is that NASA has decided to terminate the Space Launch Initiative, including the Next Generation Launch Technology Program and a number of advanced rocket engine R&D programs have been terminated. And the funding intended for hypersonics R&D is being shifted to the exploration systems activity. As I understand it, a number of the activities undertaken in the Space Launch Initiative represented NASA's contribution to the joint NASA–DOD National Aerospace Initiative.

Last year's budget request stated that NASA's Space Launch Initiative "insures America's superiority on the space frontier in both conventional rocket and air-breathing hypersonics technology fields." And it also cited, as accomplishments, that NASA had officially established the rocket-based combined cycle and X–43C projects. So what are NASA's plans for the RBCC and the X–43C projects?

Rear Admiral STEIDLE. Yes, sir. Your statement is correct, sir, but we—the Space Launch Initiative has not been terminated, it has been kind of modified and transferred into the Transportation Systems. We have taken the Orbital Space Plane and the NGLT, the Next Generation Launch Technology, and have taken the lessons learned and the concepts and the pieces of those particular programs as the starting point and the baseline for our exploration program, which is the CEV systems of systems pieces of it.

Part of the NGLT piece, as you correctly pointed out, was the hypersonics piece. I did a technology assessment of a number of programs, 140 programs total, and those different pieces were in there as well as we made the transition. There are several pieces that are going on and several of them that did not fit into where we are headed with exploration. The X–43A program is a piece of that. We did a business case assessment of: does the demonstration of those particular capabilities meet our needs in exploration? And the answer came back yes, if we have a disciplined, demonstrated performance of one flight, second flight, third flight. If we meet those objectives in each one of the flights, the Office of Exploration Systems will continue to fund that program through demonstrated performance.

We looked at NASA-unique technologies, and we made an agreement with the Office of Aeronautics that my particular Office of Exploration will fund the non-procurable pieces of that NASA-unique pieces for hypersonics study and development and that the Office of Aeronautics will fund the procurement pieces of that. So we continue on with that program in our commitment to hypersonics.

There was a piece, the X–43 program. It did not fit our needs. The X–43C, it did not fit our particular needs at this particular point for an exploration systems development program, so it was, indeed, terminated. We shared that information as we went forward with the baseline assessment and the cost benefit analysis of that particular program, and it was, indeed, canceled. We are doing that throughout all of the programs and refocusing all of our tech maturation programs as we go forward in the exploration piece of it.

Mr. LAMPSON. Is it an accurate statement to say that the funding that had been intended to follow the hypersonics work, such as the X–43C project, will be retained by the Exploration Systems Office and used to support other activities?

Rear Admiral STEIDLE. Not exactly, sir. We—our funding was reduced $130 million last year and part of finding offsets for that reduction, the X–43 fit in that particular area.

There were some other things that came out that I want to make sure I am aware of—make you aware of. In that program, also, is risk mitigation systems engineering development and some very, very fine work in the people in Marshall that we are bringing on to our program now. In fact, the director of the NGLT program, I have just selected him to be my deputy director for Transportation Systems. So besides the technology pieces of it, there is a wealth of experience in personnel that I am moving on, either actually physically moving to Washington or moving into these programs.

Mr. LAMPSON. The—what is NASA's plan for the RS–84 Reusable Rocket Engine Program?

Rear Admiral STEIDLE. Sir, that did not fit at this particular time. I canceled that program. I called the contractor personally, the President of that company, and told him that that does not fit into the needs of our program at this particular time. I had him—he came to me, Byron Wood is the CEO of that particular company, and I sat down with him and said, "This does not fit in our particular program at this time, and this is why, and this is what the business case analysis shows of that. However, Mr. Wood, I want you to participate in the program and continue on with the association with where we are headed in the future." And he has agreed, and he is pursuing that.

NASA Contribution to the National Aerospace Initiative

Mr. Lampson. What specific projects, if any, will NASA be contributing to the joint NASA–DOD National Aerospace Initiative?

Rear Admiral Steidle. There are three sections of that. On the hypersonics side, we are going to continue on to fund the X–43 demonstration through its second, and possibly, its third flight for demonstration of the Mach 10. We will fund the NASA-unique pieces of hypersonics work, and that is mainly personnel that are supporting the interfaces with the NAI. And I hope to increase the emphasis on the other two pillars of NAI, that being space access and space technology pieces of it.

Mr. Lampson. Thank you. Dr. Sega, what do you understand that NASA's future role will be in the National Aerospace Initiative?

Dr. Sega. As we formed the technology roadmaps, and that was a detailed process from a technology perspective, and it went from goals, objectives, technical challenges, approaches, and tasks, there was an understanding that they needed to fit within the programmatic requirements and strengths of the organization. So we have outlined the participation of all—of the partners and the organizations in each of these areas. Now we are not sure, at this point, how the details and the programs will continue forward by NASA in these three areas. We anticipate that the collaboration will be there where it continues to make sense. In the areas where we had anticipated NASA participation, such as the work in some of the hypersonics work, we will continue our hypersonics research and development. It will go at a bit slower pace and will have a bit of an increased risk, but our program will continue.

Mr. Lampson. Thank you.

Future of Space Launch Initiative Projects

I would like to request that NASA provide a list of all of the previously planned Space Launch Initiative projects and what NASA's intentions—what NASA intends to do with each, including the consequences for both the civil service and contractor personnel. If we could have that——

Rear Admiral Steidle. Yes, sir.

Mr. Lampson.—done at some point, I would appreciate it.

And I would yield back my time, Mr. Chairman.

Chairman Rohrabacher. Mr. Lampson, if I could make a recommendation; put that request in writing and that you copy my office, because I want to tell you that I have seen so many Members over the years make requests and it just never happens. Now I am sure that the Admiral is new to his job, and every time he gets a request from Congress like this, he is going to make sure that the answer—that the questions are answered, but I think that we ought to just have a policy from now on, whenever we are asking members of the Administration, we put it in writing and we hold people to get answers to our questions.

Mr. Lampson. I would be happy to do that, Mr. Chairman.

Chairman Rohrabacher. All right. Thank you very much.

And Mr. Feeney from Florida, who is, of course, the energetic representative from that part of Florida who serves as our launching area for America's space program. You may proceed.

Mr. FEENEY. Thank you, Mr. Chairman. And some days I have more energy than others, but as you pointed out, the Space Coast blends the NASA Civilian Launch System. We have got the commercial launch facilities, and we have got, of course, Air Force launch facilities, as well. I think, as General Dickman notes, because of your—when you handled that command of the 45th Space Wing, there are some times that the artificial boundaries that we create between those three launch facilities sometimes create some difficulties in coordinating amongst the three different launch facilities. Sometimes they have historically impeded cooperation, although we have had some successes. And I am glad that there is a reemphasis on not just the relationship between NASA and DOD, but also, obviously, NASA and the private facilities, and not just Lockheed and Boeing that have a significant presence in the Space Coast, but also Mr. Musk's and other, you know, entering facilities out there. So I am very grateful that we are having this hearing today.

NATIONAL SECURITY INTERESTS IN SPACE EXPLORATION

I wanted to ask the panel, in general, two questions and then leave it to you to address them during my time. One is with respect, specifically, to President Bush's new proposal and vision for a future journey in space, and specifically as he talks about the United States' national security interests. What specific national security interests do you think that the mission that he has laid out, the journey for the future of NASA, need to be addressed? And as part of that, General Dickman, you have mentioned in your written remarks that there are certain areas that are unsuitable for sharing between the military and NASA. If you would sort of outline what they are and why it is unsuitable to have cooperation. I understand there is a need for secrecy, for example, but if you could be more specific and help me, in my mind, clarify the appropriate places where cooperation should not be expected.

And then finally, I think Mr. Musk has a great recommendation that NASA and DOD find a way to adopt cross-certification, where applicable, so that we don't have to have redundant hurdles in getting space flight either of the manned or unmanned side, especially in the commercial area, and obviously the regulatory burden that has made it very difficult for commercial entities to—and created a barrier for greater commercial growth in space. So if the panel would sort of address those two questions during my time, I would be grateful.

Rear Admiral STEIDLE. I will start out with the vision, and we are in the process of defining our requirements. And there are not any specific requirements along those lines for national security investments and integration pieces of it, but I think the point is if—from my particular understanding of it, is perhaps if we go along, perhaps we develop our requirements in the Exploration Office for a time for rendezvous and docking capabilities or, perhaps, the remote sensing capabilities, those things may be applicable to other areas of technology maturation as well. Although we don't have anything specifically entailed, the collaboration is necessary so that we share these particular technology areas as we go forward.

There were some examples that came out of the partnership, particularly in hyperspectral imagery. NASA had an EO–1 satellite capability, and it was almost ready to be decommissioned or not to be used when, at the partnership, the exchange of information on the requirements showed that some other agency needed that capability, and it was passed on to that. So if we redeem this collaboration and we define the requirements to share our tech maturation programs, I think we will come up to some of those, although they aren't defined right from the start.

Mr. FEENEY. Well, also the infrastructure sharing, like the EELV, for example, and the capabilities.

Major General DICKMAN. I think, Mr. Feeney, there is going to be an enormous benefit for the Department of Defense and the national security space community from the increased emphasis on space and the related technologies that will come from the exploration. Whether it is bringing young people through high school and college educations that are more focused on engineering to the specific technical base itself, we will benefit directly from that flow, that increased emphasis, that will be more like where we were in the 1960s and the 1970s than where we have been over the last 10 years.

With respect to your specific question of where I think it would be inappropriate to share, the first, and most obvious, is where there is a direct application of technologies where the technology level we may be sharing to a weapon system. I refer, additionally, for example, the thermal protection systems where, at the basic materials level, we will work very closely with NASA, but as we extend that technology to, for example, nose cones on ICBMs, that is not NASA's business. It would be totally inappropriate for NASA to be involved in designing a nuclear weapon delivery system. That is what our job is. As we become more reliant on space in wartime, and certainly in Operation Iraqi Freedom. That was clear, not only to us, but to our adversaries. We will be far more concerned about both defending our own space assets and denying space capabilities to others if they choose to use them to attack the United States or threaten our interests. That is also not NASA's business, that is our business. And while we may have common science and technology, the translation of that into systems is one that is the responsibility of the Department, and not a shared collaboration with NASA.

Third, and perhaps more vision in the future, but more real now, is military presence and a base on the Moon. It is prohibited by treaty. It is not our business to be doing that. It is NASA's challenge now from the President to go to Mars and go to the Moon, and so while we will assist as best we can at the launch pad and whatnot, there will be no military bases on the Moon. Those are three examples of where that translation from the shared programs to uniquely military or uniquely NASA, I think, are appropriate.

Dr. SEGA. I would—the segment about—what General Dickman said about the impact on the national security interests, with advancing technology, you gain benefits. You gain benefits from civil systems, from military systems. And the excitement and motivation for folks to pursue science and engineering education and leading to a more robust aerospace world would be a positive outcome of the increased excitement in this general area.

NASA AND DOD CROSS-CERTIFICATION OF LAUNCH VEHICLES

Rear Admiral STEIDLE. Karen has expertise in the cross-certification piece.

Ms. PONIATOWSKI. Yeah. Mr. Feeney, what we have been doing, as a matter of fact, is working very closely between NASA and the DOD. Vehicles, such as the EELVs, we have a "one government" team that looks at the RD–180——

Mr. MUSK. I beg your pardon, again. I am sorry. I can't hear you at all.

Ms. PONIATOWSKI. I am sorry, Elon. Is that better?

Mr. MUSK. Yeah, that is much, much better.

Ms. PONIATOWSKI. Okay. I am sorry. I apologize.

What we have done is, in the case of the RD–180 and the RL–10, we have a one government approach where we are all read into and look at the same data at the same time. There are two parts to the certification: one is the generic how do you understand the vehicle, how do you understand how the systems work, and when failures happen, the correction of those; the other is that, for each mission you have, you go through a launch review to make sure that the mission overall, the changes you might have made and the vehicle you are actually flying, that it is going to have the highest probability of success. And so there are two different things. For when you own a mission and you are flying it, there is one set of certification that you do for flight readiness for that given mission, and then there is a more fleet-wide, consistent type of a work. And what we have been doing with the Air Force, in a very close partnership across all of the various vehicles, is working together on understanding that baseline understanding and anomaly resolution, so as we come to each individual launch, we are partnering off of what we have learned across the board.

Chairman ROHRABACHER. And Elon, would you like to contribute to this part of the questioning?

Mr. MUSK. You know, I apologize, cross-certification?

Chairman ROHRABACHER. We are talking about cross-certification.

Mr. MUSK. Yes, and I apologize, I only heard a portion of Karen's response, but I—you know, I think—I am sorry. I have to get—unfortunately, I just didn't hear the—most of the response, but it seems to me that it would make sense to, as much as possible, not duplicate efforts, and I assume Karen is. That sentiment makes sense. Yeah. Sorry.

Chairman ROHRABACHER. It is okay.

DIFFERENCE BETWEEN MILITARY AND CIVIL SPACECRAFT

The Chairman will now reclaim the time. Thank you. And I—things that Elon didn't hear, General, what percentage of the payloads that the military puts up are weapons systems? I mean, most—aren't these satellite sensors and a lot of things that are very similar to civilian payloads?

Major General DICKMAN. Certainly there are, and most of the technologies would apply. The weapons that we put on the front end of ICBMs are, obviously, very different.

Chairman ROHRABACHER. And there is very—we are putting very few nuclear weapons on to the heads of ICBMs these days, aren't we? I mean, this is——

Major General DICKMAN. There are still 500 warheads that——

Chairman ROHRABACHER. Yeah, but I hope we are not putting any—putting them on any new rockets. I don't think that that would be——

Major General DICKMAN. Oh, that is correct.

Chairman ROHRABACHER. Okay. So in terms of what we are doing now in the future, there is certainly a huge crossover between what you are doing in space and what NASA is doing in space in terms of sensors, satellites, you have observation satellites, communication satellites, et cetera, et cetera.

Major General DICKMAN. I am sorry, Mr. Chairman, if I suggested anything other than that. It is clear that in science and technology, many of the systems and subsystems that we work on are very common.

Chairman ROHRABACHER. I am just—this is a guess, I would say the vast majority. I am just guessing now, being a big supporter of the SDI and Missile Defense, I know that we don't have that based up there yet, but I would hope someday we do. But seeing that it is not up there yet, the huge number of the satellites, military satellites, are very similar to the civilian satellites that are put into orbit. And if that is the case, wouldn't just making sure that we have sort of a consistency of requirements rather than—and an attempt to be consistent within the requirements, wouldn't that be something that makes common sense?

Major General DICKMAN. I think that for the requirement for technology, that is probably correct. But NASA doesn't have the responsibility, for example, for tracking ICBM or missile launches that take place either in Iraq or from Russia or somewhere else. So as we take the technology for IR detectors and translate it into a Space-Based Infrared System, it is a different endgame than NASA would in taking those same sensor technologies and building an IR telescope to look into space.

CREW EXPLORATION VEHICLE

Chairman ROHRABACHER. Yeah, but we are going to—I am going to get to that in my very last question, which is going to be about this near-Earth object that just came by and how—which fits right into your answer there, but first I am going to ask a couple more questions. And I believe Dr. Sega was talking about the first flight of the crew exploration vehicle. Was that your testimony? Was that you, General? Admiral?

Rear Admiral STEIDLE. It must have been mine, sir, okay.

Chairman ROHRABACHER. Yes, it must have and because I just read it in my notes here. You said the first flight of this crew exploration vehicle is going to be in 2014?

Major General DICKMAN. Yes, sir, the first manned flight of that, that is correct.

Chairman ROHRABACHER. Holy cow. That is 10 years from now.

Rear Admiral STEIDLE. Yes, sir.

Chairman ROHRABACHER. Can't we do anything, you know—does it take 10 years to build something?

Rear Admiral STEIDLE. Well——

Chairman ROHRABACHER. When it took 10 years to get to the Moon, we built everything, and actually had the mission to the Moon and back in 10 years—less than 10 years.

Rear Admiral STEIDLE. It could be, yes, sir. What I—what we have got out there is a demonstration program in '08, which is very well achievable, not necessarily an orbital flight, but a demonstration of systems, so it is much more than the vehicle itself, it is the systems that go with it and the protocols and those pieces. And then we follow that up about 2½ years later in the start of 2011 with an orbital demonstration of that capability, docking, autonomous rendezvous capabilities. Do we need to do in-space assembly? So there are a lot of systems that go along with this as well. And of course, what we are looking at is the exploration piece, not just a one data point, perhaps, a capsule, but the development of a whole program or systems of systems. And that is what we will be developing when we put this up into orbit with humans in 2014.

Chairman ROHRABACHER. What about it, Elon? Do you think the private sector could build something, if they had this kind of budget, in less than 10 years?

Mr. MUSK. I would say without a doubt. I mean, one has to establish, sort of, the private sector in smaller entrepreneurial companies versus the——

Chairman ROHRABACHER. The large companies that are——

Mr. MUSK.—large aerospace giants, like Boeing and Lockheed, which I think the fundamental issue with the pace of progress is not so much—it is not a question of NASA being to blame. I think I would place, really, the bulk of the blame on the options that NASA has available to it from industry. And I think that there may be some new and better options from SpaceX and, perhaps, from other companies.

Chairman ROHRABACHER. Well, I am just noting that we have the, you know, Dick Rutan, out there about ready to show us what can be done in terms of a suborbital spacecraft, and I just have a feeling that the more we get small companies into this and into the mix that there is going to be a lot of—let us put it this way. There is going to be a transfer of know-how and technology from the private sector into the public sector as compared to what we used to have where all of it was the military and the public sector providing technology for people in the private sector. And I find—Elon, I find a 10-year timeline to be just—look, I am out of here; I won't be around here in 10 years. Is anybody in this room going to be around here in 10 years?

Rear Admiral STEIDLE. Sir, can I——

Chairman ROHRABACHER. Sure.

Rear Admiral STEIDLE. Yes, sir. Because what we are looking for is something beyond just a spacecraft. It is the system of systems that is going to go and do something that we have never done before. It is going to go beyond. The Moon is going to be a piece of it, but it is going to be on. It is going to be a spiral development program. It is going to be systems to support that. It is going to be—in that particular time frame, we will be going to the Moon and leaving an orbiting communications satellite.

Chairman ROHRABACHER. Right.

Rear Admiral STEIDLE. It is the development of our infrastructure. So there is quite a bit. Now if those particular requirements were just to produce a vehicle and put it in orbit, we could do that, but to be the right vehicle to grow into something for an exploration vision, that takes an awful lot of work. Now——

Chairman ROHRABACHER. Well, let me suggest that if we end up, and I knew that when the President made his vision statement that one of the Achilles' Heels is that if people start thinking that they are developing something right now that is going to be used on the Mars mission for the humans to go into Mars, it is going to create a lot of waste, and it is going to drain huge sums of money. Mr. Lampson is already afraid about the money that is being drained away from other programs. And you start talking about developing the craft for the Mars program now, there is not going to be any money left for anything. And I—when you—it seems to me, we are talking about some crew transportation vehicle that should have something to do with achieving the goal on the Moon, and then we are going to find out what we need for Mars. But if it is taking 10 years because we have got to take all of those other things into consideration, no wonder it is taking 10 years. And I will tell you, it is going to be expensive. How much are you suggesting this crew exploration vehicle will cost, as it stands now in a 10-year program?

Rear Admiral STEIDLE. We don't have the end pieces of that, sir, but what we have is from here to '09 and '10 time frame and the development of that and all of the demonstration programs to go and——

Chairman ROHRABACHER. And how much is that?

Rear Admiral STEIDLE. That is about $6.8 billion that is in the—in our program that——

Chairman ROHRABACHER. Holy cow. That is over a—that is $6 billion over a five-year period, right? Okay.

Rear Admiral STEIDLE. Sure.

Chairman ROHRABACHER. And then—but we are not going to get anything that we really are going to be using. I mean, there is a demonstrator, but we won't be using——

Rear Admiral STEIDLE. No.

Chairman ROHRABACHER.—it for another 10 years?

Rear Admiral STEIDLE. No, sir. If we do this right—and that is what a spiral development program is. If we do this right and set the requirements to where we eventually want to go so we don't eliminate and don't have some false starts and don't start down this path for something in '11 and then have to start over in '14 and develop this program so that we can have a vehicle in '14 that can be adapted just a couple years later to go to the Moon and back and support those particular pieces, that is how much it costs to be able to do a detailed program like that.

Cooperation in Near-Earth Objects Detection

Chairman ROHRABACHER. Okay. One last question, and then I will turn it over to Mr. Lampson. About the near-Earth objects, about these near-Earth objects, one, as I say, I announced at the beginning of this hearing, there is a near-Earth object that came within 25,000 or 26,000 miles of the Earth. And it is passing by tonight, I guess, or today sometime. We didn't know about it intil four days ago. I consider—now this particular one isn't big enough to have caused major damage, but the fact that we really didn't know about it until a few days ago indicates that there could be something out there that might actually be a threat to the world. Maybe we could just go down the panel and you could suggest to me what you think would be a good way for NASA, the Department of Defense, and the private sector to work together in terms of meeting this potential challenge.

Rear Admiral STEIDLE. I can't answer that question fully, sir. I could tell you it is a very small object, 30 meters, that is 43 kilometers away will go down through the Atlantic Ocean in—about—I have the time is about 17:08 tonight, but probably late. But that is a—I am going to pass it on to my experts here who can say that sounds like a very difficult one to track. We picked it up on Monday evening. Lincoln labs did a—It was put on our web at JPL on Monday night.

Chairman ROHRABACHER. Now who discovered it?

Rear Admiral STEIDLE. I was informed that Lincoln labs did, sir.

Chairman ROHRABACHER. Okay.

Major General DICKMAN. I am not familiar with how Lincoln discovered it. Obviously an asteroid is a passive object, and so tracking an asteroid is—requires either an optical system or—well, it requires, essentially, an optical system or radar, if it gets much closer. But losing radar 1/r4, you can't get very far away to do a serious radar track. And so we provide our optical sensors to NASA——

Chairman ROHRABACHER. We have got a radar station on the Moon.

Major General DICKMAN. It is still a long—it is still 1/r4 to wherever the asteroid is. Radar, we don't lose much distance going through the atmosphere, so I am not sure that would help us a whole lot, although it is certainly worth looking into.

Chairman ROHRABACHER. All right. See, I don't—I certainly plead guilty to not being a technical expert on these things.

How can you work with NASA more effectively in this challenge?

Major General DICKMAN. Right now, we provide our optical sensors to NASA for tracking those—tracking asteroids in space. The best sensors we have are optical sensors that are maintained by the Department of Defense by our Directed Energy Directorate. And I think we provide something like 18 nights a month every month to NASA for doing that tracking.

Chairman ROHRABACHER. Dr. Sega.

Dr. SEGA. Our role, in the Department of Defense, has been principally the support of the—NASA's Near-Earth Asteroid Tracking program. And so it is in the area of, as Bob said, the optics, the processing, and so forth of datas where the technology that is either developed or there is additional sensors that can be supportive of the NASA tracking program would be the role of the Department.

Chairman ROHRABACHER. And as I mentioned to you earlier, Doctor, I went to a Shuttle launch, and the Shuttle went off—took off, and we were all standing there, and within two minutes after the Shuttle took off, a comet went directly over the Florida station there at Cape Kennedy. And everybody was just aghast, because, I mean, no one had any idea that this thing was coming, but it just—it was close. It was very close. And it seems to me that NASA and the military need to look at this as a potential threat. Look at the Moon. Those craters didn't just get there by themselves. These aren't volcanic craters.

And Mr. Musk, are you ready to go and to recruit the oil workers off of the derricks in order to fly up to the asteroid and destroy the asteroid before it gets to the Earth? That is what was in the—that was in the movie, I think.

Mr. MUSK. Yes, sir. It is ironic what they have accomplished in movies. You know, the fact of the matter is that the state of technology today: if the asteroid is big enough for us to see it, we won't be able to stop it, and if it is small enough—if it is so small that we don't see it, we won't get it in time. That is sort of where we are today. Some decades on, that, hopefully, will all be changed. Certainly, of all of the threats humanity faces over the long-term, some sort of asteroid, either at—can either significantly damage civilization or possibly end it. It is highly probable that that possibility certainly exists, and I hope that, at some point in the coming decades, we are able to present a reasonable counter to that threat.

Chairman ROHRABACHER. All right. Well, thank you. And I think the private sector will play a role in that. And if by nothing else, keeping the cost of the rockets down that are part of the system.

Thank you very much.

I would like to turn to Mr. Lampson. Would you——

Mr. LAMPSON. I would. You know, one thing that we could do, Mr. Chairman, is to have—get a greater look at the points contained within the Space Exploration Act, because it does, indeed, set a goal of building a craft that would allow us to go out and do something to those asteroids when and if they are discovered. And we are going to have to address it. You are—you know, this is a serious issue, and it is something that we ought to be talking about. So hopefully we will at some point in time.

ROLE OF INDUSTRY

As much as I respect and appreciate not just Elon Musk's company, but all of them, I think it important—we don't have the opportunity for Boeing and Lockheed Martin to be here to make some comment in reference back to some comments that were made about their size and how slow they might be in accomplishing something, but they are doing things that this government has asked to do that perhaps adds cost to a lot of what they do, because they have to maintain a lot of personnel to accomplish something over a long period of time they might not otherwise be doing. I think that is important to note at some point.

Crew Exploration Vehicle

And I—Admiral Steidle, I wanted to also get you to just make a short comment about, perhaps—that you were talking about all of the complaints of the development of the CEV and what all we have to go through in order to meet a 10-year accomplishment for flight. What is the difference in what we did with Apollo during those—all of those Apollo years to develop all of the craft that were involved in those missions?

Rear Admiral STEIDLE. That particular program was a—I would call a single data point. That was one particular event and a destination that we were flying to, and all of our efforts and integration efforts that—were focused on the development of that particular system to accomplish that. This program differs in the fact that we are looking beyond that. We are looking at a continuum. We are looking at an infrastructure that will, as well as just the spacecraft, support future missions. We are looking at a spacecraft that we will slowly develop, through spiral development, to take us into orbit, to take us to the Moon, to take us beyond into journeys to, perhaps, Mars and beyond. So there is much more into it than that single data point; there is the infrastructure. And I keep referring to this, and we heard it earlier, the systems of systems integration piece of this is quite large, not just the development of the spacecraft itself, but all of the supporting structure, the support of the infrastructure that will be here on Earth, the demonstration of capabilities either on the Space Station or on the surface of the Moon and the other things that come together to support this development. So this is the significant difference from that.

Mr. LAMPSON. There is also the role that money has played in it, too. Was there not a little different——

Rear Admiral STEIDLE. Yes, sir. If you look at the funding profile in that particular program, it was a program manager's joy. It was——

Impediments to NASA–DOD Cooperation

Mr. LAMPSON. Okay. Let me get back to what I wanted to ask about in the first place. I would like to ask all of you, if you could, to respond to this. What do you consider to be the biggest impediments to meaningful cooperation between NASA and DOD in space transportation, and how would you overcome those impediments? And try to keep them as succinct as possible. Would you start, Admiral?

Rear Admiral STEIDLE. Yes, sir. Very, very quickly. I have looked at the collaboration efforts that are taking place. We mentioned the partnership. We mentioned NAI. I asked my staff to put together where do we stand and what kind of collaboration efforts are underway, and they came with a long list. And there are almost 400 relationships and collaboration agreements and MOUs between our centers and our departments in NASA. There are some main program pieces that are working here.

What I have seen, as I have gone back and studied this in preparation for coming here, that something changed about two years ago. The emphasis came from the top down, not from the bottom up. This is my own assessment. It appears that something took place at the partnership, perhaps just the new people that are involved in that. But their focus on the collaboration and the partnership was felt throughout the organization, all of the way down to the tech maturation people and the people managing each program. So the impediment would be the loss of that particular dedication and focus on making these things happen.

Mr. LAMPSON. Thank you. General.

Major General DICKMAN. I think the greatest impediment in a long-term collaboration is that we have different requirements for the vehicles when they finally get put to use. There is no impediment to collaboration in developing the technology, and the work that Admiral Steidle mentioned was specifically focused on the whole spectrum of technologies with—related to launch. The collaboration is very, very good.

With respect to the comment that the partnership changed two years ago, I would share that view as well. And since Dr. Sega might not want to comment since he was one of the players in that, the reality is, the membership of the Partnership Council changed. It includes the Administrator of NASA. The Administrator is very proactive about dealing with the Department of Defense. It includes Under Secretary Teets, but it also includes Dr. Sega and Admiral Ellis and General Lord from Space Command, who are all committed to making the partnership work, rather than getting together and seeing where there might be opportunities for interface.

Mr. LAMPSON. Dr. Sega, do you want to make a comment?

Dr. SEGA. Sure. The—I think the environment is very positive, not only in the Space Partnership Council, but also at the working level as we worked hard over a period of over a year to develop these technology roadmaps. Now once we have this in place, the challenge is to keep the integrated program on track. We have participation from the Army, Navy, and Air Force in developing some of the technologies with DARPA and, of course, the Department of Defense, and NASA inside the government. We have industry, and so forth, engaged. We also have some of the budgets, as they come here, are viewed here in Congress. It is important that when there are connections being made, that that information be passed to you so that there is an understanding of where the integration is, where the connections are as one looks at the programs in total versus in isolation, so keeping it together all of the way through the process.

Mr. LAMPSON. Mr. Chairman, if you will indulge me and let me—and allow me to ask just a couple more questions, I will be—I will pass this.

Chairman ROHRABACHER. Go ahead quickly.

Mr. LAMPSON. Okay. I will be quick.

NATIONAL SPACE TRANSPORTATION POLICY

The Bush Administration—and this is, again, for everyone. The Bush Administration has indicated that it is reviewing the existing national space transportation policy. Have any of you played a role in that policy review?

Major General DICKMAN. I have, Mr. Lampson.

Mr. LAMPSON. Can you—is it Admiral? Have you or Dr. Sega?

Rear Admiral STEIDLE. No, sir. I am too new to be a part of that. I believe Karen's office has, though.

Mr. LAMPSON. All right. And Dr. Sega?

Dr. SEGA. Not at this time.

Mr. LAMPSON. What sorts of issues are being considered in that review, and when do you expect the review will be completed, both of you?

Major General DICKMAN. We are optimistic that the review will be completed by summer. The issues are, if you will, the ones you would expect: what kind of work should be done in terms of next generation launch vehicles, in terms of innovative approaches for a workforce for launch vehicle, in terms of what NASA would need for exploration, for what kind of expendable launch infrastructure we should sustain over the next decade and beyond. Really, it is a very comprehensive look at how we are going to do space transportation.

Mr. LAMPSON. Karen?

Ms. PONIATOWSKI. Yeah. I think the only thing I would add to that is it does reinforce the need for the—unlike the previous policy that said the Air Force would have one focus and NASA another, the policy is again focused at cooperation between the NASA and the DOD.

Mr. LAMPSON. Thank you.

NASA–DOD TEST FACILITIES

And the last thing, what specific steps are NASA and DOD taking to ensure that the national test facilities needed for future space transportation systems will be maintained? We know that some of these things were—are being shut down. We know that there have been some facilities that—at NASA that haven't been maintained and are actually—we are losing the capability to provide some service with those at some point of time in the future. Are there any specific facilities that any of you are concerned about, and if so, what are they? Everybody.

Dr. SEGA. It wouldn't be fair to give any specifics, because there are an awful lot of facilities, but there are needs in terms of testing. I gave an example of the HyFly test. In this case, it was a DARPA–Navy program testing at Langley and soon to have a single-engine demonstrator test from the Air Force and DARPA also at the Langley wind tunnel. So this is a case I know personally, and it is an area that you look at what is coming forward and you look at the test facilities they need. Now in the future, technology will move forward and your needs will change, but I think it is a question that needs to be evaluated continually.

Mr. MUSK. I think, in general, the test structure right now, at least for propulsion, is fairly robust. Certainly it is robust for testing spacecraft. I believe there are five test facilities that can handle a one million-pound class engine, a variety of fuels and flows and sizes. There are test facilities that are relatively new that can handle high-altitude engine testing. So until we actually break new ground on what we need that would drive us to some place we haven't been before, I think we are probably in pretty good shape right now.

Mr. LAMPSON. Thank you.

Admiral.

Rear Admiral STEIDLE. I have been—in my initiation in NASA, I have been making trips to as many of the facilities as I possibly can, looking at their lab structure. In my previous position, I was commander of a number of centers, so I know what to look for. I have been looking for those particular capabilities. I haven't found anything lacking, so far, in my trips to them.

Mr. LAMPSON. Those are some of the things that I have heard from some NASA installations around. I went—that is more—thank you all for your answers, and thank you for your indulgence, Mr. Chairman.

Chairman ROHRABACHER. Thank you.

Mr. Feeney.

Mr. FEENEY. Thank you, Mr. Chairman.

CROSS-CERTIFICATION OF LAUNCH VEHICLES

When I last addressed you, you talked about the increasing co-operation. Specifically, if we could get a commitment to try to find some redundancies in the certification requirements, I think that would be helpful and useful to us. And if you are having success and everything, we would appreciate any continued success.

But I want to——

Chairman ROHRABACHER. Perhaps we could be very specific on that. If you could, within a month, get back to the Committee a letter indicating what areas you think that you could work on in this cross-certification area, and making sure that we don't—where it is that duplicate and where—with the military and where can people accept the standards and the—already the specifications that are on the other side that have already been proven so that people like Mr. Musk can invest $1 million and can expect it not to have to be doing the same exact thing with that $1 million with NASA that it does with the Department of Defense, I mean, rather than waste that money.

Mr. Feeney, go right ahead.

Mr. FEENEY. Well, despite the optimistic answer, Mr. Chairman, that we got, I think that there is cause for optimism. You remember Lockheed Martin that has testified in front of this committee that they think a one-government team would be very helpful to them. The *Space News* has recently reported that while these meetings are going on, very little is accomplished, and these may be accusations that are inaccurate, but if you can give us details, we, at least, can defend you, at a minimum.

REVIEW OF SPACE TREATIES AFFECTING DOD AND COMMERCIAL USE OF THE MOON

And secondly, I would like the Admiral and the General to tell me whether it would be a useful exercise for this Partnership Council, or some other group involving the civilian and the military and the commercial launch advocates in the United States, to review the 10 or 15 different space treaties that we have out there. For example, General, you suggested that one of the reasons that we do—we couldn't put a permanent facility on the Moon was not that it was Air Force related, it is because of treaties. Well, treaties come and go, and by the way, the United States tends to honor its commitments; other nations don't, many times. You are tracking some 8,000 pieces of "space junk," as we speak. It may turn out that the Moon, for example, is a better place to do some of those things or enhance what we are doing.

And finally, on the commercial side, there is very—we have used space treaties to socialize the rewards of exploring space. So if Mr. Musk, for example, wanted to find a way to colonize a square mile of the Moon, he would have to share everything that he could accomplish with 191 other nations recognized by the State Department. One wonders, in that light, whether the Queen of Spain would have subsidized Columbus' journey knowing that she would have had to share all of the fruits of his discoveries with the Dutch and the French and the Brits. So maybe you could tell me a little bit about a review of space treaties in our future.

Major General DICKMAN. We will be glad to take that as an action.

Chairman ROHRABACHER. Now with that said, we are going to adjourn now, because I do have a plane out at six o'clock at Dulles. I want to thank Mr. Lampson. This is his first hearing as the Ranking Member. Thank you very much for your—he has always been an activist and always had thought-provoking questions and different insights that he has shared with us. We are very happy now that he is playing a more important role. You are not just a Member of the Committee; now you are the Ranking Member. And someday, who knows, you may sit in this chair. So I have got to be—I have to be very, very, you know, courteous to this man.

But Mr. Feeney, thank you, as well. And thank you to our panel. I appreciate it.

One thought, we did say a couple things that seemed derogatory about big companies, because we have got Elon Musk, this great entrepreneur here who has put his own money in, and we always admire that so much, but just——

Mr. LAMPSON. We need them both.

Chairman ROHRABACHER. We need them both. And for the record, Lockheed, Boeing, Orbital, all of them were contacted about the hearing, and all are providing written testimony for the record in this hearing and have been very cooperative with us and couldn't cooperate more. They just—so we had a good hearing today, and I would like to thank the witnesses again. And please be advised that the Subcommittee Members may request additional information for the record, and I would ask other Members who are going to submit written questions to do so with one—within one week of this date. And I would then conclude this hearing by saying this meeting is now adjourned.

[Whereupon, at 3:50 p.m., the Subcommittee was adjourned.]

Appendix 1:

Answers to Post-Hearing Questions

ANSWERS TO POST-HEARING QUESTIONS

Responses by Rear Admiral (Ret.) Craig E. Steidle, Associate Administrator, Office of Space Exploration Systems, National Aeronautic and Space Administration

Questions submitted by Chairman Dana Rohrabacher

Q1. In written testimony provided to this committee, Michael Gass of Lockheed Martin Space Systems recommends that "there ought to be one government team [between NASA and the Department of Defense (DOD)] to define mission assurance requirements" for launch vehicles before entering into a development program with industry. He explains that this approach "would eliminate wasteful redundancy, use common processes for acquisition and contract oversight, share the benefits of complementary hardware systems and reduce the cost of maintaining infrastructure." (See Appendix 2: Additional Material for the Record for Mr. Gass's full testimony.)

Will NASA and the Air Force consider this "one government team" approach in defining launch vehicle requirements to meet both NASA and DOD requirements? What are the pros and cons of such a proposal for NASA?

A1. NASA, the Air Force, and the NRO are looking to maximize our limited resources in space launch. To that end, we collaborate in a number of areas. For example, we share data and common analysis sources. We participate in an inter-agency working group. We share common risk lists between organizations for our common rocket fleets. We participate in each other's Pedigree Reviews and Design Equivalency Reviews on the key engine systems in the EELV fleet.

However, the concept of a common set of mission assurance requirements, while attractive, does not recognize the different emphasis each organization places on various elements of mission assurance. Differences in requirements stem from the difference in how the systems are to be used. A common set of requirements would end up being composed of the lowest common denominator. It would treat a human space launch, a low-cost technology demonstration satellite, and a National security missile warning satellite as equals. Certain missions demand unique mission assurance investments and scrutiny that are not appropriate or affordable for all flights. We need the flexibility to succeed at this priority, rather than having a common set of requirements that we constantly deviate from and issue for waivers based upon the mission's importance.

Q2. During the course of the March 18th hearing, both you and Major General Dickman agreed to investigate how NASA and the Air Force's separate certification processes for new launch vehicles could be better coordinated. The concern was that these separate certification processes might be redundant, wasting taxpayer money. What progress has been made to-date? What plans have you made to minimize the redundancy in the certification process?

A2. NASA and the Department of Defense (DOD) collaborate on technical assessments (sometimes referred to as certification) of the launch vehicles we each use to accomplish our unique space missions. Common reviews and technical assessments on both existing and evolutionary systems have moved us toward a better understanding of the core launch system. Each agency continues to perform mission assurance and "certify" each individual mission for launch (including the integrated spacecraft and launch vehicle as ready for launch). However, neither the United States Air Force (USAF) nor NASA formally "certify" a launch vehicle design, as is done by the Federal Aviation Administration (FAA) for commercial aircraft in order to ensure public safety.

The U.S. has not fielded many "new" national launch systems; for example the DOD did not have a use for the Athena or the Conestoga launch system and was not engaged in the technical assessments for those "new" vehicles. In the case of the Pegasus, NASA worked closely with Defense Advanced Research Projects Agency (DARPA) and the USAF to understand and address challenges met in the start up of the commercial program. NASA and DOD have built a pattern of cooperation in the sharing of common launch systems; we would expect to extend that cooperation to the fielding of "new" systems. Likewise, every individual user, be they commercial or government, employs varying levels of technical assessment leading up to a unique mission. We would expect that practice to continue on new launch systems as well.

Q3. To accomplish NASA's missions to the Moon and Mars, NASA may need to develop a new heavy-lift launch vehicle beyond the capabilities of the current

Evolved Expendable Launch Vehicles (EELV). At a recent Senate Appropriations hearing, NASA Administrator Sean O'Keefe said there are "competing options and alternatives" for heavy-lift launch vehicles using the Space Shuttle stack of boosters and external tank or enhancing DOD-developed expendable launch vehicles.

What is the timeframe for NASA's decision-making on whether to develop and how to develop a heavy-lift launch vehicle?

A3. NASA intends to make a decision on both cargo and human launch capability by the release of the CEV Level 1 requirements in January 2005.

Q3a. What trade studies are NASA conducting to answer the question on whether or not a heavy-lift launch vehicle is needed? Why wouldn't existing or slightly modified expendable launch vehicles be sufficient for NASA's space exploration initiative?

A3a. NASA has initiated trade studies that consider options using Space Shuttle propulsion elements, commercial systems, international systems, and Evolved Expendable Launch Vehicles (EELV). The study teams are considering payload mass capabilities from 25mT (the current capability of the EELV heavy) up to 100mT. The trade parameters currently focus on performance, non-recurring cost, and recurring cost. A trade study has also been initiated that will define the human rating requirements for a CEV and launch system. These trade studies will be completed this summer. The follow-on trade studies will combine the CEV human launch capability and cargo launch system requirements into a common study that will evaluate the synergy between the two system requirements. NASA's objective is to separate crew and cargo launch to the maximum extent possible. Industry has also been encouraged to consider these critical launch trades as they prepare responses to the Request for Information (RFI) and proposals to the Broad Agency Announcement (BAA) for Concept Exploration and Refinement.

Q3b. Why aren't the designs for NASA's Saturn V rocket adequate for NASA's future heavy-lift launch needs?

A3b. Elements of the Saturn launch system are being considered in the trade studies for cargo launch, including the J–2/J–2S engine that provided propulsion for the Saturn upper stage and the Apollo trans-lunar injection stage. The J–2/J–2S powerhead was most recently used to power the X–33 aerospike engine that was successfully demonstrated. Other elements of the Saturn launch system may be applicable to a modern launch capability. Although the Saturn launch vehicle design would provide the performance capability to conduct a lunar mission, the capability to manufacture it does not exist today and would be extremely expensive to duplicate. More recent launch systems have incorporated modern manufacturing processes and structural materials. Building from the current launch vehicle manufacturing and launch operations infrastructure may be more affordable than resurrecting the Saturn; but the trade studies are assessing all possible launch solutions.

Q4. Both NASA and the DOD are reporting to Congress a 50 percent cost increase in what industry plans to charge for future purchases of Evolved Expendable Launch Vehicles due to a sharp downturn in the commercial launch market and U.S. launch providers not being able to recoup their costs for these new rockets. Could the U.S. Government possibly achieve some cost savings if NASA and the DOD bargained together when negotiating its launch vehicle purchase prices with industry? If so, are there any plans to begin such cooperative bargaining?

A4. NASA and the DOD have a practice of coordinating on launch vehicle acquisitions. We have established forums and working groups to assure that our respective launch requirements and acquisition plans are discussed to maximize the overall national benefit. NASA and the USAF closely coordinated our individual requests for proposals for recent awards to Boeing for Delta II activity. We have also shared our requirements for EELV class services to facilitate assurance that the U.S. Government is getting a fair and reasonable price. The depressed international market for U.S. launch services in the EELV class has indeed raised concerns with the potential for cost increases to all government users. We are working with the DOD to understand and address these concerns.

Q5. NASA recently informed the Committee that it is no longer planning to fly secondary research payloads on the Space Shuttle, affecting 125 experiments. The European Ariane rocket has developed a fairly robust capability to carry secondary research satellites into orbit, while the U.S. has not developed an analogous means to carry secondary satellites on an EELV. Are NASA and the DOD developing an analogous capability for U.S. expendable launch vehicles? If not, why not? Could some of these secondary payloads be launched on an Ariane rocket instead?

A5. NASA has invested funds to enable secondary capability on existing launch systems and has flown approximately 15 secondary payloads over the past 10–15 years using Delta, Pegasus and Taurus launch systems. With the advent of the EELV, the USAF has invested in the development of the EELV Secondary Payload Adapter (ESPA), which is scheduled for flight in 2006. Flight opportunities for small payloads have been a topic of discussion between the agencies. In December 2003, NASA, the National Reconnaissance Office (NRO) and the USAF met to discuss both the opportunities for excess space on future EELV missions and investment strategies for complementary secondary carriers beyond ESPA. All agencies have been actively engaged in maximizing any excess space and performance on larger launch systems and will continue to coordinate efforts.

The majority of the experiments/payloads that will no longer be flown on the Space Shuttle are Get Away Special (GAS) payloads that were expressly designed to fly as tertiary payloads on a "space-available" basis. Most were designed to require human intervention and/or return, which are unique Shuttle capabilities, and are not readily converted for flight on expendable systems.

Q6. A recent news article said that cost overruns with the Defense Advanced Research Projects Agency's Orbital Express satellite project threaten to shut down the project. NASA invested $25 million in DARPA's Orbital Express project to demonstrate robotic docking technology applicable to future missions to the Space Station.

If DARPA decides to cancel the Orbital Express project, what would the impact be on NASA's investment and technical objectives?

Is DARPA coordinating its decisions with NASA on the Orbital Express project?

A6. NASA has invested $17 million in DARPA's Orbital Express project and intends to invest another $8 million through FY 2004 and FY 2005. The Orbital Express takes the next step beyond NASA's Demonstration of Autonomous Rendezvous Technology (DART) flight project in demonstrating robotic docking technology. Orbital Express will utilize the Advanced Video Guidance System (AVGS) that will be demonstrated on DART to maneuver within close proximity to the target satellite but, unlike DART, Orbital Express will demonstrate actual docking with the target satellite. Both experiments demonstrate technologies that may be used in the on-orbit assembly of elements of an exploration mission. If DARPA decides to cancel the Orbital Express project, NASA would incorporate that decision into the technology gap analysis that is currently underway. NASA will assess the risk to the overall concept of operations and develop a risk mitigation plan to reduce or eliminate that risk. NASA supports the Orbital Express project and believes that the partnership with DARPA is the most effective way to reduce the risk of autonomous rendezvous and docking.

NASA is currently providing engineering support to the Orbital Express project. DARPA has coordinated their project plans with NASA and intends to complete the project.

Questions submitted by Representative Nick Lampson

Q1. Please provide a list of all of the previously planned Space Launch Initiative projects, their intended five-year funding profiles, the numbers of civil service and contractor personnel involved in each, and what NASA intends to do with each project, including the consequences for both the civil service and contractor personnel.

A1. See Enclosure 1 for a list of Space Launch Initiative projects.

Space Launch Initiative projects

Projects	Civil Service FTE	Support Contractor FTE	Prime Contractor FTE	FY04 Initial Operating Plan $M	FY05-FY09 Total Planned	Status
OSP Total	**420**	**266**	**562**	**382**		
OSP Vehicle	361.4	231	448	320		Contracts not extended further
DART	43.6	24	79	27		Continue
PAD	15	11	35	27		Option not exercised
Orbital Express				8		Continue
NGLT Total	**702**	**454**	**648.5**	**388**	**1036.3**	
RBCC	12	2	37	8	0	Discontinued
TBCC	41	14	69	23	78	Discontinued
X43C	71	20	50	34	215	Discontinued
Propulsion Technology	205	131	287.5	119	467.3	Discontinued*
Vehicle Systems	173	78	166	101	143	Discontinued**
Systems Engineering	200	106	39	84	133	Continue
Interdisciplinary				13		Continue
Earmarks				6		NASA is assessing
X37	118	75	432	177		Continue

* The Integrated Powerhead Demonstrator (IPD), Auxiliary Propulsion, and materials and sensor tasks will continue. *Civil Service employees will work continuing project, transfer to Exploration Tasks, or transfer to other programs. University Research and Engineering Technology Institutes will continue.*

** The Vehicle Subsystems power and electric actuators will continue.

Enclosure 1

Q2. In your announcement to Congress regarding the cancellation of the X–43C contract, you stated: "It is important to note that the FY 2004 NGLT budget reduction will only impact contracted activities. NASA civil service workforce efforts will continue in-house.. . ." *This statement leaves the impression that NASA is not concerned with the impact that the abrupt cancellations have on contractors, and is only focused on protecting its civil service workforce. NASA officials have indicated that NASA's Office of Aeronautics is considering whether it might be able to pick up some of the hypersonics work that has been terminated. If there is the possibility that NASA will seek to restart some hypersonics activities in the FY 2006 budget request, what steps are you taking now to minimize the disruption to the contractor workforce between now and FY 2006?*

A2. NASA is always concerned about its contractor teams and any adverse impacts Agency decisions may have on them. However, part of the rationale for outsourcing an activity includes the increased flexibility that NASA gains for later reprogramming or termination of a program/project. This flexibility is lost when NASA performs the work in-house based upon the Government civil service laws. Consequently, X–43C contractors were allowed time to close out the contracts in an orderly way, including placement of the employees connected to the contract. NASA is currently reviewing the options to ease the transition as it considers the priorities of future hypersonics research. Should NASA validate the requirement to restart the hypersonic research in FY 2006, the FY 2005 funding will be re-examined, and if funds are necessary, NASA will work with Congress to adjust the FY 2005 operating plan to ensure an orderly transition. It is also noted that the Department of Defense continues its hypersonics activities (e.g., the single engine demonstrator project) using many of the same contractors.

Q3. What specific factors led you to decide to continue the X–37 program? What data or systems do you expect to come from that program that will help you achieve the goals set forth in the new exploration initiative? Please be specific. How much will the X–37 program cost in total, and what is included in that cost estimate?

A3. The X–37 project provides hardware development experience for the NASA/industry team that will be applied to the exploration mission. The Exploration Mission requires development of new technologies to achieve the mission objectives and the X–37 effort provides key technology development in thermal protection systems and hot structures.

The X–37 Approach and Landing Test Vehicle (ALTV) effort will provide an in-flight calibration of air data system technologies, and verification of the aerodynamics and guidance and control for a vehicle other than the Space Shuttle. The data obtained will be utilized to validate analytical models and trade studies. The orbital vehicle technologies (wing leading edge thermal protection system and hot structures) will provide higher reentry temperature capability to apply to the exploration mission. The hot structures development will develop capability in the U.S. that currently only exists overseas.

The estimate of total X–37 program cost is currently being developed and will include contractor effort, government in-line effort, government insight and management.

Q4. What role, if any, did DOD play in the decision to continue the X–37 program? Will DOD provide any funding for it?

A4. In the summer of 2001, the Air Force decided to complete its commitment to the X–37 project of $16 million but provided no additional funding. This commitment was completed in FY 2002. NASA decided independently to continue the X–37 project in 2003 via the NRA 8–30 competitive process. DOD was not involved in that decision. NASA encourages potential partnerships where appropriate, however, no DOD funding is currently committed.

Q5. NASA and DOD are collaborating on the National Aerospace Initiative (NAI). However, it is hard to get a clear picture of the extent of the cooperation. Please provide for the record: a list of the projects that make up the NAI, which of them are jointly funded, and the estimated cost of each of the NAI projects.

A5. The following is a list of NAI projects that will continue:

The Integrated Powerhead Demonstrator (IPD) will continue from the NGLT Propulsion Technology. NASA and the Air Force Research Laboratory jointly manage IPD. NASA will provide $3.4 million in FY 2004 and will provide $7 million in FY 2005. NASA is also providing 25 civil service and 27 support contractors to the effort. IPD is on the NAI Space Access Roadmap and is managed by the DOD Integrated High Payoff Rocket Technology Program.

The X–43A project will continue through the third flight in FY 2005. X–43A is a critical step in the development of scramjet technology and is on the NAI Hypersonics roadmap.

Other continuing projects will contribute to the NAI Space Access roadmap, including Auxiliary Propulsion, Vehicle Subsystem power and electric actuator technology, and the University Research and Engineering Technology Institutes. NASA is providing funding of $25 million in FY 2004 and $21 million in FY 2005. NASA civil service and support contractor employees supporting these projects in FY 2004 is 121 and 98 in FY 2005.

Q6. NASA's decision in the 1990s not to be the first user of an unproven rocket was based on a series of failures of three new launch vehicles (the Pegasus XL, the Conestoga, and the LLV). How should the government weigh the risks versus the benefits when considering the use of new launch vehicles? That is, where should the government draw the line between prudent stewardship of the taxpayers' money by not risking a taxpayer-funded satellite on an unproven launch vehicle versus the need to encourage the development of new launch vehicles that could ultimately reduce governmental launch costs?

A6. NASA has developed a risk mitigation process that seeks to balance mission criticality with flight history. The policy enables NASA to launch on a first flight of a brand new system, the second flight (post demonstration flight), or systems with proven demonstrated flight history. Over the past several years, NASA has not identified any missions that can tolerate the risk of a new launch system. NASA payloads, even the small payloads, tend to be unique, one-of-a-kind efforts, with a higher overall mission cost.

NASA does believe in encouraging the development of new launch vehicles and emerging launch companies. NASA is partnering with DARPA on their FALCON Program, which will provide flight demonstrations for new launch systems. The payloads will be more risk tolerant and better suited to initial flights of new launch systems. Through this partnership, the government agencies are each able to participate in enabling new launch systems with a balanced risk approach.

ANSWERS TO POST-HEARING QUESTIONS

Responses by Major General (Ret.) Robert S. Dickman, Deputy for Military Space, Office of the Under Secretary of the Air Force, Department of Defense

Questions submitted by Chairman Dana Rohrabacher

ONE GOVERNMENT TEAM

Q1. In written testimony provided to this committee, Michael Gass of Lockheed Martin Space Systems recommends that "there ought to be one government team (between NASA and the DOD) to define mission assurance requirements" for launch vehicles. He explains that this approach "would eliminate wasteful redundancy, use common processes for acquisition and contract oversight, share the benefits of complementary hardware systems and reduce the cost of maintaining infrastructure." (See Appendix 2: Additional Material for the Record for Mr. Gass's full testimony.)

Will NASA and the Air Force consider this "one government team" approach in defining launch vehicle requirements to meet both NASA and DOD requirements? What are the pros and cons of such a proposal for the Air Force?

A1. All agencies—the Air Force, the NRO, and NASA—are looking to maximize our limited resources in space launch. To ensure best use of these resources, we collaborate in a number of areas. For example, we share data, common analysis sources, and have coordination meetings. We share common risk lists between organizations for our common rocket fleets. We participate in each other's Pedigree Reviews and Design Equivalency Reviews on the key engine systems in the EELV fleet. The Air Force and the prime contractors are looking for ways to develop common support in order to reduce the program costs. However, common acquisition processes would likely add complexity and cost.

The concept of a common set mission assurance requirements, while attractive, does not recognize the different emphasis each organization places on various elements of mission assurance. A common set of requirements would end up being composed of the lowest common denominator. It would treat a manned space launch, a low cost technology demonstration satellite, and a national security missile warning satellite as equals. Our history is not all space capabilities are equal. Certain missions demand unique mission assurance investments and scrutiny that are not appropriate or affordable for all satellite flights. Mission assurance, often referred to as mission success, is the #1 priority for AF, NRO, and NASA space programs. We need the flexibility to succeed at this priority, rather than having a common set of requirements that we constantly deviate from and waiver based upon the mission's importance.

SEPARATE CERTIFICATION PROCESSES

Q2. During the course of the March 18th hearing, both you and Rear Admiral Steidle agreed to investigate how NASA and the Air Force's separate certification processes for new launch vehicles could be better coordinated. The concern was that these separate certification processes might be redundant, wasting taxpayer money. What progress has been made to-date? What plans have you made to minimize the redundancy in the certification process?

A2. Neither the Air Force nor NASA formally "certify" a launch vehicle design, as is done by the FAA for aircraft. The Air Force and NASA do share information to gain understanding of flight worthiness that aids in certification of the booster. Technical teams share information on the pedigree of flight hardware—for example, Air Force and NASA teams review each other's hardware on Delta II rockets, in addition to hardware for commercial missions, and share their findings to assure they understand the state of the fleet. Mission teams discuss common rehearsed goals and techniques between Air Force and NASA Mission Directors. Air Force and NASA teams jointly work test requirements for hardware—for example, the new solid rocket booster on the Atlas V will be tested to a program defined by the contractor, NASA, and the Air Force. Additionally, the Air Force, National Reconnaissance Office, and NASA hold a mission assurance forum in which the contractors and government teams look for synergies, best practices, as well as lessons learned. The most recent forum was successfully concluded this past March of 2004.

HEAVY-LIFT LAUNCH VEHICLE

Q3. If NASA needs to develop a heavy-lift launch vehicle for its space exploration missions, do you foresee possible uses for such a launch vehicle for future DOD missions?

A3. The recent Air Force mission model (from 2000–2020) does not require development of a heavy-lift vehicle beyond the current capability of the EELV fleet. NASA's heavy-lift requirements and architecture appear to be still developing. It is unclear where the NASA heavy-lift requirements, once defined, can be satisfied by the existing heavy-lift capability of our EELV fleet. However, NASA is keeping DOD informed of any contemplated performance or reliability enhancements to the EELV fleet to allow DOD to consider how those potential enhancements would benefit DOD mission needs.

EELV

Q4. Both NASA and DOD are reporting to Congress a 50 percent cost increase in what industry plans to charge for future purchases of Evolved Expendable Launch Vehicles due to a sharp downturn in the commercial launch market and U.S. launch providers not being able to recoup their costs for these new rockets. Do NASA and the DOD coordinate their launch purchases before going out with bids to industry? Could the U.S. Government possibly achieve some cost savings is NASA and the DOD bargained together when negotiating its launch vehicle purchase prices with industry?

A4. Although NASA and the Air Force do not coordinate with each other on their respective launch purchases, they do share appropriate information. Despite requirement differences between NASA and the Air Force, it is possible there could be some synergy if NASA and DOD bargained together for EELV launch services; however, whether this would yield any saving over the current process is unknown.

Questions submitted by Representative Nick Lampson

HUMAN-CARRYING SPACE TRANSPORTATION SYSTEM

Q1. Do you foresee a DOD requirement for a human-carrying space transportation system?

If so, when?

What would be the reason for the requirement?

Will DOD either review or participate in the formulation of the requirements for NASA's proposed Crew Exploration Vehicle?

A1. There is no Air Force requirement for a human-carrying space transportation system at this time. However, the Air Force intends to support NASA's effort to formulate requirements for their Crew Exploration Vehicle (CEV), especially as it relates to EELV.

These questions were submitted to the witness, but were not responded to by the time of publication.

Q2. NASA's decision in the 1990s not to be the first user of an unproven rocket was based on a series of failures of three new launch vehicles (the Pegasus XL, the Conestoga, and the LLV). How should the government weigh the risks versus the benefits when considering the use of new launch vehicles? That is, where should the government draw the line between prudent stewardship of the taxpayers' money by not risking a taxpayer-funded satellite on an unproven launch vehicle versus the need to encourage the development of new launch vehicles that could ultimately reduce governmental launch costs?

ANSWERS TO POST-HEARING QUESTIONS

Responses by The Honorable Ronald M. Sega, Director, Defense Research and Engineering, Department of Defense

Questions submitted by Chairman Dana Rohrabacher

Q1. In the wake of the Space Shuttle Columbia *tragedy and need to finish Space Station construction, NASA recently informed the Committee that it is no longer planning to fly secondary research payloads on the Space Shuttle. According to Major General Dickman's testimony, over 200 DOD experiments have flown on over 70 Shuttle missions. NASA reports that over 125 satellite experiments, including DOD payloads, will be affected.*

Q1a. What is the impact to DOD space research if these Shuttle rides are canceled?

A1a. The reduction in science and technology demonstration opportunities would limit collection of scientific data and could result in increased program development risk. Shuttle and International Space Station flights are an effective means for developing and testing technologies, especially for those test units and experiments requiring post-flight physical analysis.

Q1b. The European Ariane rocket has the means to carry secondary research satellites into orbit. Are NASA and the DOD developing an analogous capability for U.S. expendable launch vehicles?

A1b. The Space Test Program (STP) and Air Force Research Lab (AFRL) developed the EELV Secondary Payload Adapter (ESPA) that can fly multiple space experiments. The ESPA fits any EELV medium-class launch vehicle and holds up to six small satellites, in addition to the primary payload. The first ESPA flight will be on a dedicated STP launch aboard a medium EELV in September 2006. Success of this first ESPA flight would provide DOD space experiment possibilities on future EELV medium launch vehicles.

Q2. If NASA needs to develop a heavy-lift launch vehicle for its space exploration missions, do you foresee possible uses for such a launch vehicle for future DOD missions?

A2. Launch needs for DOD and National Security missions are under the purview of the Under Secretary of the Air Force for Space.

Questions submitted by Representative Nick Lampson

Q1. What, if any, are DOD's interests in NASA's X–37 program?

- *What role, if any, did DOD play in NASA's decision to continue that program?*
- *DOD previously had withdrawn funding for the X–37 program. Will DOD provide any funding for it in the future?*

A1. The proposed X–37 Orbital Vehicle (OV) has the potential to demonstrate technologies that would have a range of utility to both DOD and NASA. The technologies that could be demonstrated on the X–37 OV include: non-toxic, storable propulsion and power; advanced thermal protection system materials and structures; materials, structures and components for long-duration exposure to the space environment; advanced flight control systems; and algorithms for autonomous in-space, reentry and landing maneuvers.

Q2. NASA and DOD are collaborating on the National Aerospace Initiative (NAI). However, it is hard to get a clear picture of the extent of the cooperation. Please provide for the record: a list of the projects that make up the NAI, which of them are jointly funded, and the estimated cost of each of the NAI projects.

A2. The National Aerospace Initiative serves an integrating role for Science and Technology (S&T) investment to focus national aerospace research and technology development and demonstrations toward goals and objectives which support future high payoff capabilities. Technology roadmaps were developed in the areas of high speed/hypersonics, access to space and space technology through a collaborative effort between DOD and NASA. The NAI and NASA's Office of Exploration Systems are currently identifying and evaluating NAI S&T projects that will support the Vision for Space Exploration. Attachment 1 contains a listing of DOD projects which include NAI S&T activities. The attachment indicates overall project funding levels and the level of NAI investment within that particular project. NAI funding is also split into two categories: core and enabling technologies. Core technologies and enabling technologies, such as materials, aerodynamics, guidance and control, and power support NAI and other S&T objectives.

Q3a. Do you foresee a DOD requirement for a human-carrying space transportation system? If so, when? What would be the reason for the requirement?

A3a. On July 22, 2002, the Deputy Commandant of the Marine Corps for Plans, Policies, and Operations published a Universal Need Statement (UNS) for Small Unit Space Transport and Insertion Capability (SUSTAIN). This UNS outlined the need to deliver 13 combat-equipped personnel through and from low earth orbit. The UNS did not specify a date required.

Q3b. Will DOD either review or participate in the formulation of the requirements for NASA's proposed Crew Exploration Vehicle?

A3b. The DOD will participate as appropriate.

Q4. NASA's decision in the 1990s not to be the first user of an unproven rocket was based on a series of failures of three new launch vehicles (the Pegasus XL, the Conestoga, and the LLV). How should the government weigh the risks versus the benefits when considering the use of new launch vehicles? That is, where should the government draw the line between prudent stewardship of the taxpayers' money by not risking a taxpayer-funded satellite on an unproven launch vehicle versus the need to encourage the development of new launch vehicles that could ultimately reduce governmental launch costs?

A4. Every launch decision must weigh a complex series of factors to determine acceptable risk. The analysis approach of weighing risk versus benefits of use of new launch vehicles should be accomplished on a case by case basis. For example, the launch of Tactical Satellite I (TacSat I), an R&D satellite, is currently scheduled this year on the first flight of a new booster called Falcon.

Attachment 1

NATIONAL AEROSPACE INITIATIVE (NAI)

NAI is based on 3 pillars: High Speed / Hypersonics, Space Access and Space Technology. Each pillar includes various projects and elements of projects. In many cases, NAI work only represents a portion of a particular project. To capture the total investment in NAI, the funding is split into 2 categories: core and enabling technologies. Core technologies directly support the level I demonstration roadmap. Enabling technologies, such as materials, aerodynamics, guidance and control, and power support both NAI and other S&T efforts.

ARMY

BA	PE	PE TITLE	BPAC	BPAC TITLE		FY04	FY05
2	0602211A	Aviation Technology	47A	Aeron & Acft Wpns Tech		$35,533	$37,716
					NAI Investment	**$1,379**	**$1,431**
2	0602211A	Aviation Technology	47B	Veh Prop & Struct Tech		$3,630	$3,913
					NAI Investment	**$1,800**	**$2,063**
2	0602303A	Missile Technology	G02	NAI Applied Research		$8,130	$16,839
		(Hydrogen Scramjet, Attack/Intercept Missile Demonstrator)			***NAI Investment***	**$8,130**	**$16,839**
2	0602303A	Missile Technology	223	Aero-Propulsion Technology		$24,301	$0
		(Hydrogen Scramjet, Attack/Intercept Missile Demonstrator)			***NAI Investment***	**$24,301**	**$0**
2	0602782A	Command, Control, Communications Technology				$780	$670
					NAI Investment	**$780**	**$670**
3	0603003A	Aviation Technology	447	Acft Demo Engines		$6,832	$7,157
					NAI Investment	**$200**	**$0**
3	0603006A	Command, Control, Communications Advanced Technology				$10,000	$10,400
					NAI Investment	**$10,000**	**$10,400**
3	0603008A	Electronic Warfare Advanced Technology				$2,750	$0
					NAI Investment	**$2,750**	**$0**
3	0603313A	Missile & Rocket Advanced Technology	G03	NAI Advanced Tech		$0	$0
		(Hydrogen Scramjet, Attack/Intercept Missile Demonstrator)			***NAI Investment***	**$0**	**$0**
				Army Total Investment = Core + Enabling Technologies		**$49,340**	**$31,403**

NAVY

BA	PE	PE TITLE	BPAC	BPAC TITLE		FY04	FY05
2	0602114N	Power Projection Applied Research	HSP&AWT	High Speed Propulsion & Advanced Weapon Technologies		$142,626	$98,831
		(RATTLRS, IHPTET, VAATE, HyFly)			***NAI Investment***	**$9,769**	**$29,430**

BA	PE	PE TITLE	BPAC	BPAC TITLE		FY04	FY05
3	0603114N	Power Projection Advanced Technology (RATTLRS, IHPTET, VAATE, HyFly)	R2911	Power Projection Advanced Technology		$198,188	$86,179
					NAI Investment	**$32,680**	**$41,955**
3	0603236N	Warfighter Sustainment Advanced Technology (IHPTET, VAATE)	R2915	Warfighter Sustainment Advanced Technology		$52,531	$61,103
					NAI Investment	**$10,000**	**$10,000**
	TBD				***NAI Investment***	**$8,100**	**$8,100**
				Navy Total Investment = Core + Enabling Technologies		**$60,549**	**$89,485**

AIR FORCE

BA	PE	PE TITLE	BPAC	BPAC TITLE		FY04	FY05
1	0601102F	Defense Research Sciences	2302	Solid Mechanics & Structures		$11,641	$13,276
					NAI Investment	**$2,478**	**$6,357**
1	0601102F	Defense Research Sciences	2303	Chemistry		$27,939	$29,292
					NAI Investment	**$11,654**	**$13,523**
1	0601102F	Defense Research Sciences	2304	Mathematical & Computer Sciences		$29,293	$25,663
					NAI Investment	**$7,937**	**$10,675**
1	0601102F	Defense Research Sciences	2306	Materials		$15,035	$15,917
					NAI Investment	**$15,035**	**$15,917**
1	0601102F	Defense Research Sciences	2307	Fluid Mechanics		$12,875	$10,902
					NAI Investment	**$7,906**	**$4,192**
1	0601102F	Defense Research Sciences	2308	Propulsion		$15,660	$15,864
					NAI Investment	**$9,648**	**$11,157**
1	0601102F	Defense Research Sciences	4113	External Research Programs Interface		$7,364	$9,638
					NAI Investment	**$750**	**$766**
1	0601102F	Defense Research Sciences			***NAI Investment***	**$4,599**	**$5,329**
2	0602102F	Materials	4347	Materials for Structures, Propulsion, & Subsystems		$64,131	$41,057
					NAI Investment	**$26,897**	**$27,429**
2	0602201F	Aerospace Vehicle Technologies	2401	Structures		$28,679	$32,831
					NAI Investment	**$8,030**	**$14,525**
2	0602201F	Aerospace Vehicle Tech	2403	Flight Controls and Pilot-Vehicle Interface		$15,486	$16,643
					NAI Investment	**$4,861**	**$6,482**
2	0602201F	Aerospace Vehicle Tech	2404	Aeromechanics and Integration		$20,146	$25,205
					NAI Investment	**$0**	**$7,296**

2	0602201F	Aerospace Vehicle Technology			***NAI Investment***	**$37,410**	**$33,384**
2	0602203F	Aerospace Propulsion	3145	Aerospace Power Technology		$35,162	$24,946
					NAI Investment	**$7,177**	**$7,539**
2	0602203F	Aerospace Propulsion	3012	Advanced Propulsion Technology		$13,790	$12,211
					NAI Investment	**$13,790**	**$12,211**
2	0602203F	Aerospace Propulsion (IHPTET, VAATE)	3048	Fuels & Lubrication		$16,612	$12,841
					NAI Investment	**$10,765**	**$10,213**
2	0602203F	Aerospace Propulsion (IHPTET, VAATE)	3066	Turbine Engine Technology		$36,533	$31,749
					NAI Investment	**$36,382**	**$30,640**
2	0602203F	Aerospace Propulsion	3145	Aerospace Power Technology		$35,162	$24,946
					NAI Investment	**$4,660**	**$2,870**
2	0602500F	Multi-Disciplinary Space Tech	5025	Space Materials Development		$19,447	$21,499
					NAI Investment	**$16,366**	**$19,355**
2	0602500F	Multi-Disciplinary Space Tech	5026	Rocket Propulsion Component Tech		$51,909	$40,961
					NAI Investment	**$35,983**	**$35,790**
2	0602500F	Multi-Disciplinary Space Technology	5027	High Speed Airbreathing Propulsion Technology		$4,549	$180
					NAI Investment	**$4,549**	**$180**
2	0602500F	Multi-Disciplinary Space Tech	5030	Applied Space Access Vehicle Tech		$0	$0
					NAI Investment	**$0**	**$0**
2	0602500F	Multi-Disciplinary Space Technology			***NAI Investment***	**$40,761**	**$34,282**
2	0602601F	Space Technology			***NAI Investment***	**$47,876**	**$38,704**
2	0602805F	Dual Use Science & Technology	4770	Dual Use Science & Technology		$10,496	$5,151
					NAI Investment	**$5,277**	**$2,589**
3	0603112F	Advanced Materials for Weapon Systems	3946	Materials Transition		$23,876	$5,298
					NAI Investment	**$22,292**	**$4,968**
3	0603211F	Aerospace Technology Dev/Demo	4920	Flight Vehicle Tech Integration		$32,075	$24,463
					NAI Investment	**$7,613**	**$7,628**
3	0603211F	Aerospace Technology Dev/Demo			***NAI Investment***	**$17,790**	**$11,863**
3	0603211F	Aerospace Technology Dev/Demo	5099	National Aerospace Initiative		$0	$0

					NAI Investment	**$0**	**$0**
3	0603216F	Aerospace Propulsion & Power Technology (SED, IHPTET, VAATE)	2480	Aerospace Fuels & Atmospheric Propulsion		$3,101	$374
					NAI Investment	**$1,699**	**$367**
3	0603216F	Aerospace Propulsion & Power Technology	3035	Aerospace Power Technology		$4,185	$4,297
					NAI Investment	**$3,471**	**$3,119**
3	0603216F	Aerospace Propulsion & Power Technology (IHPTET, VAATE)	4921	Aircraft Propulsion Subsystems Integration		$28,600	$16,719
					NAI Investment	**$28,600**	**$16,719**
3	0603216F	Aerospace Propulsion & Power Technology (SED)	5098	Advanced Aerospace Propulsion		$15,750	$26,300
					NAI Investment	**$15,750**	**$26,300**
3	0603216F	Aerospace Propulsion & Power Technology (IHPTET, VAATE)	681B	Advanced Turbine Engine Gas Generator		$29,050	$26,185
					NAI Investment	**$25,896**	**$23,366**
3	0603401F	Advanced Spacecraft Tech	682J	Spacecraft Vehicles		$25,372	$8,420
					NAI Investment	**$5,463**	**$2,491**
3	0603401F	Advanced Spacecraft Technology			**NAI Investment**	**$72,320**	**$41,402**
3	0603500F	Multi-Disciplinary Adv Dev Space Tec	5032	Advanced Space Materials		$11,615	$0
					NAI Investment	**$3,280**	**$0**
3	0603500F	Multi-Disciplinary Adv Dev Space Tec	5033	Rocket Propulsion Demonstration		$22,032	$22,437
					NAI Investment	**$15,020**	**$11,536**
3	0603500F	Multi-Disciplinary Advanced Space Technology	5062	Advanced Structures for Space Vehicles		$2,975	$0
					NAI Investment	**$2,975**	**$0**
3	0603500F	Multi-Disciplinary Advanced Space Technology			**NAI Investment**	**$32,763**	**$33,110**
4	0604855F	Operationally Responsive Launch	A013	Small Launch Vehicle		$25,844	$35,362
					NAI Investment	**$25,844**	**$35,362**
4	0604856F	Common Aero Vehicle (FALCON)	A012	Common Aero Vehicle		$12,200	$21,700
					NAI Investment	**$12,200**	**$21,700**
					Air Force Total Investment = Core + Enabling Technologies	**$653,755**	**$591,334**

DARPA

BA	PE	PE TITLE	BPAC	BPAC TITLE		FY04	FY05
2	0602702E	Tactical Technology (HyFly)	TT-03	Naval Warfare Technology		$47,613	$46,813
					NAI Investment	**$22,099**	**$19,615**

2	0602702E	Tactical Technology (SED)	TT-07	Aeronautics Technology		$16,229	$41,760
					NAI Investment	**$0**	**$15,000**
3	0603285E	Advanced Aerospace Systems (FALCON)	ASP-02	Space Programs & Technology		$201,584	$249,220
					NAI Investment	**$71,703**	**$97,375**
3	0603285E	Advanced Aerospace Systems			***NAI Investment***	**$152,282**	**$184,471**
				DARPA Total Investment = Core + Enabling Technologies		**$246,084**	**$316,461**

DARPA

BA	PE	PE TITLE	BPAC	BPAC TITLE	FY04	FY05
3	0603750D8	Advanced Concept Technology Demonstrations			$0	$1,200
				ACTD Total Investment = Core + Enabling Technologies	**$0**	**$1,200**
				TOTAL NAI Investment	**$1,009,728**	**$1,029,883**

Summary of Basic Documents and Related Materials Being Returned

OSD-Legislative Affairs:	Classification:	OSDLA-Control No:	Date Received:
Document Control Record	Unclassified	20040060	June 3, 2004

Committee Number:	Hearing Dates:
1	March 18, 2004

Committee:
House Science Committee

Subject:
NASA-DoD Cooperation in Space Transportation

Witness
No

OPR:	Committee Suspense Date:	Close Out Date:
OSD/LA	June 10, 2004	September 30, 2004

Document Type:	Page Inventory:	Document Description:
Questions for the Record		09/30/04 - Sent Transcript and Dr. Sega's QFRs to cmte.

Attention: Hammond Tom
Location: B-374 RHOB

Receiving Officer Signature:____________________ Date:__________

Answers to Post-Hearing Questions

Responses by Elon Musk, Chief Executive Officer, Space Exploration Technologies

Questions submitted by Chairman Dana Rohrabacher

Q1. Your testimony mentioned some of your concerns with the regulatory certification process and costs attendant with those regulations.

Q1a. What specific examples do you have where Air Force or FAA launch regulations were overly burdensome and not adding value in terms of safety?

A1a. To date, SpaceX has only undergone the Air Force range safety approval, but will be seeking an FAA license shortly. Our company can therefore only comment on the Air Force process, but would be happy to provide similar comments on the FAA at a later date. Most of the FAA regulations are drawn from the Air Force document EWR–127–1 and so some of the comments are likely to be common.

The Air Force and the Office of the Secretary of Defense have been working with SpaceX to improve the certification process and in bring a new launch vehicle to flight (moreover, using only thrust termination, rather than explosive termination). My recommendations below should be seen in the context of a relationship that is working well.

Specific recommendations are:

1. **Remove the piece parts & traceability requirement for flight termination systems in favor of more extensive testing**
   Piece parts and traceability requirements necessarily result in extremely expensive components, because modern (extremely reliable) electronics are mass manufactured and do not lend themselves to being built in tiny batches. If standard electronics are suitable for a flight critical autopilot on a 400 passenger 747 landing in zero visibility over a densely populated city, then they should be suitable for a launch vehicle carrying no-one departing over unpopulated ocean.
2. **Automatic safety cross-certification between the eastern and western ranges**
   At present, there are slightly different subjective preferences held by the eastern and western ranges. This requires a company to have either both ranges present throughout the EWR tailoring process, increasing expenses to both the company and the government, or work with just one and then be forced to modify the system later for launch on the other range. New airplane companies are not forced to work with FAA branches from all corners of the country and this should not be the case with rockets either. If a vehicle and its flight termination system are suitable in California, they should be automatically allowed anywhere else in the country with a similarly unpopulated flight path.
3. **Require transparency of evaluation metrics**
   One of the critical evaluation metrics is the maximum probable loss calculation for property and casualty. However, unlike the situation with FAA aircraft certification, this calculation is not done working with the launch vehicle company and the methodology is not shared. This makes it very difficult for a launch vehicle company to determine how to improve the MPL.

Q1b. How much does it cost your company per launch to comply with the Government's regulations? Of that amount, how much does your company spend to comply with regulations that you would categorize as overly burdensome and not adding value in terms of safety?

A1b.

SpaceX estimates for compliance with Government regulations:

- $400,000 for purchase, integration and installation of the flight termination system. This number was over $300,000 higher prior to being granted a waiver of the flight termination explosives requirement by Air Force range safety.
- $650,000 for range services *required* for launch, including flight analysis, flight termination system oversight, ground safety and launch oversight.

SpaceX estimates of costs that add minimal value and could be eliminated:

- $200,000 could be eliminated from the flight termination system by being able to use flight termination components with extensive testing as a substitute for piece parts and traceability requirements. Simply being able to use the low cost, high reliability flight termination receivers that are used on missiles and UAVs would net an immediate $80,000 saving. In fact, SpaceX is approved to use these receivers at the Reagan Test Site at Kwajalein, but not at the Eastern or Western ranges.

- At least $300,000 or more than half the range services cost could probably be eliminated by streamlining the launch process. Please note that SpaceX expects to yield close to this number simply by working with the ranges and increasing their comfort level with the Falcon launch system. I will not have any recommendations until SpaceX is past its second or third launch.

Questions submitted by Representative Nick Lampson

Q1. NASA's decision in the 1990s not to be the first user of an unproven rocket was based on a series of failures of three new launch vehicles (the Pegasus XL, the Conestoga and LLV). How should the Government weigh the risks versus the benefits when considering the use of new launch vehicles? That is, where should the Government draw the line between prudent stewardship of the taxpayer's money by not risking a taxpayer-funded satellite on an unproven launch vehicle versus the need to encourage the development of new launch vehicles that could ultimately reduce Governmental launch costs?

A1. Prudence is certainly warranted before placing a valuable satellite on a new launch vehicle. The chance of failure is significant, as suggested by history and common sense. However, we must necessarily venture forth and encourage new developments or be trapped into using the same transport system forever. As I will show below numerically, staying with the old transport system will very quickly overwhelm the cost of a failure or two on a new launch vehicle.

It is one thing to ask a new company to underwrite all the costs of development, it is another to force them to fly first with other customers, it is yet another to insist that serious dialogue on purchase cannot even begin until a successful launch takes place. This latter point is particularly harsh, since payload manifests are planned years in advance. If a company cannot enter serious manifesting discussions until after first successful launch, it means that first launch of a NASA payload will only occur four to five years after the maiden launch. Our company was told precisely this by the Office of Space Flight at NASA.

As it is, the Office of the Secretary of Defense has chosen to purchase the first launch of Falcon I (at a discount to the standard price), for which SpaceX is very appreciative. The second flight of Falcon I has been purchased by the Malaysian Space Agency, raising the question of why a foreign government's space agency is more supportive of a U.S. launch company than our own.

The first flight of Falcon V, our medium lift vehicle, has also been purchased, in this case by a U.S. commercial customer.

The analysis below compares Falcon I with Pegasus, the primary NASA launch vehicle for small payloads.

Falcon I versus Pegasus

	Falcon I	Pegasus
Launch Cost (all inclusive)	$6.6 million ($700k incl. as an average for range & integration costs) to place up to 1500 lb in LEO	**$30 million (NASA catalog price) to place up to 1000 lb in 110nm LEO**
Small Satellite Cost	$5 million to $50 million	**$5 million to $50 million**
Total Mission Cost (min)	$11.6 million	**$35 million**
Total Mission Cost (max)	**$56.6 million**	**$80 million**

As can be seen from the above table, at the low end of small satellites, it would make more financial sense to buy three complete satellites, launch them on Falcon I and have two fail completely, than buy one launch of Pegasus ($34.8M versus $35M). The numbers become even more compelling if you consider that SpaceX offers discounts for purchases of three or more flights.

However, even considering the high end for small satellites, it makes sense to choose Falcon I. Using the commercial launch insurance market as a gauge for probable risk, the premium for a first launch is approximately 30 percent, obviously mostly for replacement of the satellite. Adding 30 percent to the maximum mission cost of $56.6M for launch on Falcon I, we obtain an amount of $73.6M, which is still significantly less than the $80M mission cost for launch on Pegasus.

Note, in this analysis I have assigned a 100 percent success probability to Pegasus. In actual fact, the failure rate for Pegasus historically has been 10 percent, which means roughly $3.5M to $8M must be added as risk premium to that price for launch. Moreover, the above analysis considers only the first few launches. A net present value of cost savings should be obtained by discounted back over the entire launch manifest, which would include dozens of launches. The value equation is therefore overwhelmingly in favor of a new, low cost rocket.

Seen in this light, the Pentagon purchase of first flight on Falcon I is not irrational or based on an appetite for more risk (even if it should turn out that our first launch fails). The same is true of the commercial customer that has bought the first launch of Falcon V. While there may be other strategic reasons for these actions, they can also be viewed simply as sound financial judgment.

Appendix 2:

Additional Material for the Record

STATEMENT OF MICHAEL C. GASS

MICHAEL C. GASS
VICE PRESIDENT, SPACE TRANSPORTATION, SPACE SYSTEMS COMPANY
LOCKHEED MARTIN CORPORATION

Mr. Chairman, Thank you for the opportunity to provide a statement for the Subcommittee hearing record on NASA–DOD cooperation. The subject of cooperation between the Department of Defense and NASA is both timely and delicate. It is timely because we as a nation are about to embark on a new mission of space exploration. All of our skills and resources must be brought to bear if this mission is to realize its ambitious goals. NASA–DOD cooperation is a delicate subject because it evokes institutional and philosophical biases that have in the past gotten in the way of mission objectives.

Lockheed Martin has, for many years, worked with each of these fine organizations. In partnership with NASA we have built spacecraft and systems that have surveyed the surface of Venus, monitored the Earth's environment, landed on Mars, photographed storms on Jupiter, analyzed the rings of Saturn and sampled the dust of a distant comet.

With DOD, we have built the space-borne eyes and ears of our military forces, from surveillance to communications to weather analysis and more. For both institutions, we have provided the boost vehicles that take these spacecraft to orbit and beyond.

Our experience with each organization has been characterized by mutual respect and a shared sense of accomplishment. We have seen over the years the thing that they and we have most in common: a dedication to mission success.

If the Space Exploration initiative is to be successful, NASA and DOD must work together. Lockheed Martin supports productive cooperation between them, focused on areas of common interest and respectful of their differing charters.

In the past, the differences between the two—in both size and mission—has caused some justifiable caution on the part of NASA. This caution is shared no doubt by Members and staff of this subcommittee. Nonetheless, there are areas in which these differences can be used to NASA's advantage, and space launch is one of those areas.

Currently, access to space for NASA missions involves four distinct sets of infrastructure: STS, Delta II, Delta IV, and Atlas V. There are good, historical reasons that this is so, but it is an expensive and inefficient way to operate.

The latter two boosters, the Delta IV and Atlas V, are new systems developed by DOD under the EELV program. Together, they constitute a robust, modular and reliable foundation for virtually all future space launch requirements, be they scientific, military or commercial. STS and Delta II have been workhorse systems for NASA. If it is to both afford and execute the critical missions of the Space Exploration initiative, NASA must rationalize and streamline its space launch infrastructure. This should be done in accordance with a strict timetable in order to achieve the necessary savings and meet the necessary milestones.

Thank you for the opportunity to respond to the four areas that will be addressed in your hearing today.

1. To what extent can NASA and the DOD benefit from greater cooperation in the development and purchase of launch vehicles?

NASA and DOD will obtain both cost and reliability benefits from greater cooperation in space launch vehicles. Launch vehicle cost and reliability are significantly driven by the simplicity of the system design and launch rate. The Atlas V program has made significant reductions in system complexity and labor by the evolutionary (spiral) development of the Nation's launch vehicle fleet. One of the main objectives of the EELV acquisition was to reduce launch costs by at least 25 percent. This has been accomplished with a combination of reduced infrastructure and improved designs. Infrastructure was reduced by going from nine heritage (three Delta, three Atlas and three Titan IV) launch pads to four pads (one Atlas and Delta pad per coast). On the vehicle side, the Atlas program has reduced the number of propulsion elements from nine on the Atlas HAS to only two on the Atlas V 401 configuration improving both reliability and increasing performance. The Atlas launch operations crew size and processing time have been reduced by 50 percent through increased use of automation, while performing 70 consecutive successful launches. If NASA were to launch science and exploration missions on EELV exclusively, the nearly doubled launch rates would minimize the need for USG-funded, fixed infrastructure, while maintaining two viable systems to provide assured access to space. This will create incentives for innovation while maintaining the benefits of competition.

Reliability Enhancements—As part of the Atlas V evolution, we have significantly improved reliability by both reducing parts and adding fault tolerance. Withy support from NASA, we have performed an initial Atlas V assessment to identify high value reliability launch vehicle improvements. The NASA study included single-point failure identification, fault avoidance and fault tolerance, design enhancement and ultimately a recommendation of the top 5–10 investments in reliability improvement. The key elements identified were the upper stage and booster engine single-point failure elimination, additional robustness of the solid rocket boosters (SRBs) and avionics upgrades. These NASA reliability improvements are currently unfunded. The AF Assured Access to Space (AATS) program is initially funding RL–10 upper stage engine producibility enhancements and critical component engineering which inherently improves reliability. Cooperative NASA and DOD funding of these high value reliability improvements would directly benefit all launch vehicle customers.

The recommended NASA approach for human rating combines highly reliable EELVs with intact crew abort systems enabled by a robust launch vehicle health management (LVHM) system. The LVHM system provides the critical sensing and indications required to initiate a crew system abort. This same LVHM sensing system would also provide a valuable Mission Assurance tool for DOD Missions by providing in flight measurements of critical systems, valuable for post flight mission assessments and feedback for future flights. This valuable activity is also currently unfunded.

Performance Enhancements—An evolutionary launch vehicle strategy allows incremental and affordable performance improvements, with demonstrated performance at each step, which benefits both NASA and DOD. This enables incremental funding decisions based upon the ultimate architecture needs. This strategy maintains common launch vehicle elements to maximize production rate benefits. It minimizes changes to the launch infrastructure (pads, vehicle integration facilities). This enables common launch infrastructure for both current and future NASA and DOD missions.

Increased launch rate provides both reduced cost and increased reliability. Common elements in the EELV configurations, with appropriate production and launch rate, maintain proficiency of the production and launch crews, resulting in increased reliability. The NASA human-rating approach will further improve reliability. Both DOD and NASA mutually benefit from launch vehicle reliability improvements.

In summary, we believe that there ought to be one government team to define mission assurance requirements. This would eliminate wasteful redundancy, use common processes for acquisition and contract oversight, share the benefits of complementary hardware systems and reduce the cost of maintaining infrastructure.

2. What steps NASA is taking to collaborate with the DOD in order to realize those benefits?

There are several positive examples of NASA/DOD cooperation:

First is the USG Partnership Council. This Council consists of the senior leadership from NASA/DOD/AF/NRO/DARPA and meets quarterly to continue to foster interagency cooperation in technology and program support.

Second, NASA is working with DOD to adapt the EELV program for the Crew Exploration Vehicle and other space launch requirements. NASA has initiated studies for EELV reliability, human rating and crew escape systems technology that will benefit all USG agency users.

Third, the NASA Prometheus initiative builds on the Navy's vast experience in nuclear power development. By doing so, it will accelerate the availability of advanced space power and propulsion technologies for civil space applications.

Finally, the National Aerospace Initiative has been focused on the development of third-generation technologies in support of hypersonics, space transportation, and in-space technologies. A U.S. government joint project office (staffed by experts from NASA, DOD and NRO) has been working with industry partners for the past several years on focused technology demonstrations in these disciplines.

Government working groups should give priority to the establishment of common launch vehicle requirements. This process would benefit both NASA and DOD missions in the future.

3. What areas of launch vehicle development are exclusively the role and responsibility of one agency or the other?

Current U.S. National Space Transportation policy establishes NASA as "lead agency for technology development and demonstration for next generation reusable space transportation systems." DOD's role in the current Policy is focused on expendable launch systems. The practical effect of this language has been to limit unnecessarily DOD's involvement in decisions pertaining to next generation reusable systems. To meet emerging national security requirements for space control and force projection, DOD should be able to fully explore next generation reusable system solutions. Conversely, this delineation has precluded NASA innovation in exploiting the DOD investment in expendable launch vehicles. As mentioned earlier, the current and foreseeable launch rates do not support the development of fully reusable launch systems.

The National Space Transportation Policy (currently in revision) should clearly state that the Department of Defense/U.S. Air Force has the flexibility to develop and utilize more responsive launch vehicle capabilities as required to support its mission requirements. The updated policy should promote full cooperation between NASA and DOD on both next generation reusable and expendable space transportation systems and clearly articulate the desirability of comprehensive collaboration.

4. To what extent can NASA and the DOD encourage the growth of the U.S. domestic launch market, including emerging U.S. launch vehicle providers who provide unique capabilities?

Due to the collapse of the commercial launch market and relatively flat demand for government launches, the current domestic launch vehicle providers are at a greater than 50 percent over capacity. This has driven prices to below cost, an unsustainable condition.

By concentrating both NASA and DOD space launch demand around the two EELV vehicles, the government can help stabilize and strengthen the industrial base on which space access depends.

As regards emerging U.S. launch vehicle providers, the newly-identified need for "responsive" launch capabilities may provide an opportunity. DARPA, for example, is developing the Force Application Launched from CONUS (FALCON). This and related technology programs, driven by military utility, offer the best prospect for these entrepreneurial business. They would be unwise to project a commercial demand that covered their cost.

Many, including Lockheed Martin, have developed small launch vehicles only to find that the market is not adequate to support the cost. Selling at below cost to establish a market has not proven to be a successful long-term strategy. Emerging launch vehicle providers face the historic challenge to make a viable business case in an unforgiving environment.

Summary

There is an old adage among pilots that says, "Plan your flight and then fly your plan." With this subcommittee's oversight and support, NASA is being reinvigorated and refocused. If it is to achieve its new goals, however, NASA needs to stick to a well-thought-out plan. A key enabler is assured access to space. This plan must include a reduction in the complexity, cost and management burden of its current launch infrastructure. In cooperation with DOD, NASA can take advantage of an adaptive, responsive range of boosters to meet the needs of its exciting future.

STATEMENT OF NORTHROP GRUMMAN

NASA–DOD Cooperation in Space Transportation

Mr. Chairman and Members of the Committee, thank you for convening a hearing on March 18, 2004, on the important topic of cooperation between NASA and the Department of Defense on space transportation. Northrop Grumman believes this is an important issue, and we would like to take this opportunity to provide input to the process.

Developing safer, more reliable and cost-effective space transportation systems is the key to all future U.S. activities in space. New systems will mean more robust capabilities to defend America and improve our national security. New launch capabilities are a prerequisite to turn our dreams of exploring the Moon, Mars and beyond into reality, because without new systems we will never achieve those dreams. Such systems would support viable opportunities for industry in space and enable us to explore concepts such as generating energy from orbit. Science missions would also benefit since launch costs—often half the cost of a science mission—would be less significant.

The Defense Department needs operationally responsive spacelift—vehicles that can be launched on a few hours' notice, from a variety of locations and with adaptable launch facilities. The current systems, although technologically impressive, are slow and cumbersome, and they cost so much to operate that they preclude our nation from taking the best advantage of innovative thinking on national security from space.

NASA needs a system that augments safety while decreasing costs, especially if it is to achieve sufficient cost savings to support its bold new mission to explore the Moon, Mars and beyond. As Robert Heinlein once said, "Reach Earth orbit and you're halfway to anywhere in the solar system."

These two agencies may seem to be driven by different requirements, yet they are not mutually exclusive. When there are so many pressing concerns facing our nation, we don't have the luxury of developing a vehicle that fills only a single need. A next-generation launch system can be developed that will meet the needs of both DOD and NASA.

Later this year the White House is expected to issue a new national space transportation policy. Preliminary drafts indicate that the policy will eliminate the current bifurcation that limits NASA to studying the development of reusable vehicles and DOD to studying the development of expendable vehicles. This is a positive step. There are great synergies between the needs of NASA and DOD, and although the Nation needs to be careful not to develop a vehicle that attempts to be all things to all customers, the two agencies should work together to develop a vehicle architecture that can serve both of their needs.

There are three key aspects to ensure that the United States stays at the forefront of space launch and develops the systems that will support our future national security and exploration needs:

The Government must make a long-term commitment to sustained investments in space launch technologies

A spiral development approach that makes gradual improvements is essential to continued success

NASA and DOD should reach out beyond the traditional base of launch systems providers to seek out new technologies and approaches

NASA's recent investments in the Space Launch Initiative and Next Generation Launch Technology programs have led to tremendous strides in the development of numerous launch vehicle enabling technologies like Composite Cryogenic Tanks and Integrated Vehicle Health Management Systems. Similarly, DOD's FALCON program and future investments in Operationally Responsive Space will continue to expand our knowledge and capabilities. Taken together, these NASA and DOD programs are critical investments in U.S. launch competitiveness.

These programs are an indication that our leaders are beginning to understand that space launch is not a near-term technological hurdle, but an enduring national necessity. The United States must invest in these systems and technologies on a regular and continuous basis, allowing for steady progress and continual improvements to keep this nation at the forefront. Our national security and our scientific and technological industry base require such a sustained commitment.

Using the stepping-stone, spiral development approach to space launch—moving from expendable launch vehicles, to hybrid systems that are partially reusable, and finally to fully reusable systems—is a viable way to ensure steady improvements that will keep the United States at the forefront of space access. It is time to move beyond 30- and 40-year old technologies and build the launch systems that can sustain space development safely and affordably, and in the long run reusable systems are the clear choice. To get there does not require a revolution, but regular technological improvements that build upon each other. In the same way we evolved from the Model T Ford to the safer, less expensive and more reliable automobiles of today, our nation should make the investments in launch technologies that will move us from expendable vehicles to the reliable workhorses of the future.

A number of companies have innovative solutions to U.S. space launch requirements, and both DOD and NASA should look beyond the established space launch providers. As recent work on FALCON, NASA's Alternate Access to Station, and the National Aerospace Initiative demonstrate, many of the most inventive solutions can come from companies that take a fresh approach and have not been immersed in traditional technologies and methods of operation from the past 30 years. Northrop Grumman is certain we are one of these companies that can bring critical breakthroughs and innovative thinking to our nation's space launch challenges, and we hope that space leaders recognize the importance of competition and pioneering contributions from those who are not wedded to the status quo.

Cooperation between NASA and DOD is central to successful advancement in meeting this nation's space launch requirements now and into the future. We need a government-wide assessment of requirements that establishes clear objectives and acknowledges the technical tradeoffs—that would provide the necessary direction to build a next-generation system that meets the Nation's requirements.

Northrop Grumman believes U.S. Government and industry are on the path to progress. Together we have made great technological strides in recent years, and with a sustained commitment and step-by-step spiral development we can achieve lasting success. Congress can provide the multi-year funding to continually invest in space launch technologies and oversee an obligation to an open, competitive process. With this support from Congress, industry—with the firm commitment of Northrop Grumman—will deliver the launch systems that will maintain U.S. leadership in space far into the future.

Thank you for accepting Northrop Grumman's view on this very important topic.

Prepared Statement of Wilbur Trafton

Wilbur Trafton
VP and GM, Expendable Launch Systems;
President, Boeing Launch Services

Mr. Chairman, I am pleased to offer you Boeing's Launch Services perspective on NASA–DOD cooperation in space transportation. While this is an area with a complex history, what is not often appreciated is how often NASA and DOD's space activities have resulted in synergistic benefits to each organization. This trend continues to this day.

First, however, I would like to point out two areas of cooperation that many may not be well recognized. Most specifically, in the mid-1980's, NASA's space station program, provided an excellent training ground for young space professionals who had entered the field during a lull in the industry when the Shuttle development was complete, the existing expendable launch systems were being phased out, and the DOD did not have a major new development program underway. These young professional engineers, scientists and factory workers, under the guidance of their Apollo era superiors, designed and built today's International Space Station—a technological marvel which, despite budget and programmatic issues over the years, continues to function nearly flawlessly six years after it began its on-orbit assembly in 1998.

In the mid-90's, as the U.S. Air Force sought to reduce the cost of access, the Boeing Company employed many of the same professionals, benefiting from their ISS experience, in leading the effort to design and develop the Delta IV—our entry into the Evolved Expendable Launch Vehicle (EELV) program. To date, the Delta IV has flown flawlessly three times—a tribute to these professionals and the lessons they learned while working on space station.

Now that the EELV development is complete, the U.S. has two new fleets of highly capable, modern launch vehicles which can be used both to support the original DOD mission model as well as jump start the new NASA Space Exploration Vision. While NASA is currently engaged in trade studies and analysis of the optimal space transportation solution for crewed missions to Mars and the Moon, a decade or more of preparatory, early missions can be supported by the existing Delta IV fleet and modifications to it.

Relatively straightforward and affordable upgrades to the Delta IV Heavy launch vehicle could increase our capabilities to Low Earth Orbit (LEO) from our current 23 metric tons (mt) up to 45 mt. We would encourage the USG to consider existing vehicle upgrades as an early element of spiral development to enable near-term successes for space exploration missions. This is similar to the process used in the Apollo era which grew the early Redstone and Jupiter rockets into the Saturn I and Saturn IB—which became the stepping stones which led to the Saturn V.

This spiral development approach could focus on, for example, a larger upper stage engine that would enhance Delta IV's performance to support exploration missions and could also be used to support a new super heavy-lift capability—whether an EELV-Derived, Shuttle-Derived or a Clean Sheet concept. By developing such an engine early on, the cost and schedule risk of developing the ultimate exploration launch system could be reduced and, a new in-space propulsion capability (a translunar or trans-Mars stage) could also be developed. Other technologies that could be considered include upgrades to the existing Delta IV RS–68 engine (the only new large U.S. booster engine in decades), densified cryogenic propellants, and aluminum lithium propellant tanks.

In the course of these spiral developments of the existing EELV fleets to support NASA's Space Exploration Vision, the Nation's defense will directly benefit from these upgraded performance capabilities which will be available to meet future military space requirements. And future DOD programs will leverage the engineering skill base that is sustained and enhanced in support of NASA's development efforts. In this way, synergistic cooperation between NASA and DOD will continue to benefit the respective visions of each organization as well as that of our space dependent society.

I thank you for the opportunity to submit this testimony.